Fundamentals of Horticulture: A Comprehensive Guide to Horticultural Science

Author:

Dr.Shalu Vyas

Assistant Professor
Department of Horticulture,
Sant Baba Bhag Singh University, Jalandhar
Punjab-144030

Dr. Shalu Vyas is a distinguished author and academician, possessing an impressive background in horticulture and agriculture. She completed her Ph.D. at Chaudhary Charan Singh Haryana Agricultural University (CCSHAU) in Hisar, Haryana, specializing from department of Horticulture. Her academic journey has been marked by excellence and recognition, with several notable achievements in research, teaching, and agricultural extension activities.

With a total of 10 years of experience, Dr. Shalu Vyas has demonstrated her commitment to advancing the field of horticulture through her research contributions. Her research papers have been published in prestigious national and international journals, showcasing her expertise and valuable insights in various aspects of horticulture and agricultural sciences.

In addition to her research work, Dr. Shalu Vyas has actively engaged in agricultural extension activities, through her involvement in extension activities, she has worked closely with farmers and rural communities, disseminating knowledge about modern agricultural practices, and promoting sustainable farming methods. Her efforts in this domain reflect her dedication to bridging the gap between academic research and real-world agricultural challenges.

Dr. Shalu Vyas's passion for teaching and mentoring is evident in her role as an Assistant Professor at Sant Baba Bhag Singh University in Punjab. Through her position, she imparts her extensive knowledge to young minds, nurturing the next generation of agricultural professionals and researchers. Her guidance and support have been instrumental in shaping the research interests and careers of 20 MSc students whom she has successfully guided to date.

Notably, Dr. Shalu Vyas has been honored with the Young Scientist Award in an international conference in collaboration with NITI Aayog, recognizing her exemplary contributions to the field of horticulture and research. Such prestigious recognition further reinforces her standing as an influential figure in her field.

Even during her B.Sc. program, Dr. Shalu Vyas received the Puran Anand Adhlkha Fellowship, a testament to her exceptional academic performance and dedication to her studies.

Dr. Shalu Vyas's accomplishments as an author, researcher, educator, and agricultural extensionist have left a significant impact on the scientific community and agricultural sector. Her research publications, teaching endeavors, and collaborative work with government agencies have contributed to the advancement of horticultural sciences and sustainable agricultural practices.

In conclusion, Dr. Shalu Vyas is a multifaceted professional with an outstanding academic and research record. Her journey from Chaudhary Charan Singh Haryana Agricultural University to her current role as an Assistant Professor at Sant Baba Bhag Singh University exemplifies her dedication and commitment to the field of horticulture and agriculture. Her contributions and achievements continue to inspire students, researchers, and fellow academicians alike.

PREFACE

Welcome to "Fundamentals of Horticulture: A Comprehensive Guide to Horticultural Science." This book is a journey into the captivating world of horticulture, a field that merges science, art, and nature's wonders to cultivate plants and shape landscapes.

Horticulture is a rich tapestry of knowledge that spans across disciplines such as biology, agronomy, botany, environmental science, and design. It touches every aspect of our lives, from the food we eat to the beauty that surrounds us. In this book, we embark on a systematic exploration of horticulture, unraveling its principles, practices, and profound impact on the world.

Our journey begins with the foundational concepts of plant anatomy, physiology, and genetics. We delve into the marvels of plant growth and development, uncovering the intricate mechanisms that govern everything from germination to fruiting. As we traverse the chapters, you will encounter the diverse techniques that horticulturists employ to propagate, nurture, and enhance plant species. From traditional methods to cutting-edge technologies, each approach has a role to play in the art of cultivation.

Landscapes are living canvases, shaped by both nature's hand and human creativity. Within these pages, we explore the art of designing gardens and green spaces, considering aesthetics, functionality, and sustainability. We also delve into the nuances of soil science, irrigation, pest management, and the delicate balance between maintaining plant health and protecting the environment.

Embark on this journey with curiosity and an open heart, and may the knowledge within these chapters inspire you to sow seeds of growth, beauty, and sustainability in your own corner of the world. Your valuable suggestions will go a long way to achieve this end, such suggestions will thus be warmly appreciated and heartly welcomed.

Happy gardening!

Dr. Shalu Vyas

Contents

Chapter 1:

Introduction to Horticulture

Horticulture is a diverse and essential branch of agriculture that deals with the cultivation, production, and management of fruits, vegetables, ornamental plants, and other useful crops. This chapter aims to explore the definition and scope of horticulture, delving into its historical significance and its relevance in modern society.

1. Definition of Horticulture: Horticulture can be defined as the science and art of growing and cultivating plants for human use and aesthetic purposes. It encompasses a wide range of plant species, including fruits, vegetables, herbs, flowers, medicinal plants, and ornamental plants. Horticulturists apply scientific principles to enhance plant growth, improve crop quality, and develop new varieties through breeding and biotechnological advancements.

2. Historical Evolution of Horticulture: The origins of horticulture can be traced back to ancient civilizations, where humans transitioned from a nomadic lifestyle to settled agricultural practices. Early horticulturists domesticated wild plants and selectively bred them to obtain desirable traits, leading to the development of cultivated crops. The historical significance of horticulture can be seen in the advancements of agricultural societies, botanical gardens, and the spread of plant knowledge across cultures.

3. Scope of Horticulture: The scope of horticulture is extensive and multifaceted, encompassing various disciplines and areas of application:

   a. Pomology: Focuses on the cultivation and management of fruit-bearing trees and shrubs, including apples, citrus fruits, grapes, and tropical fruits.

   b. Olericulture: Concentrates on the production of vegetables, encompassing a wide variety of crops such as tomatoes, cucumbers, lettuce, and carrots.

   c. Floriculture: Deals with the production of ornamental and flowering plants, including cut flowers, potted plants, and landscape design.

   d. Landscape Horticulture: Involves the planning, design, and maintenance of outdoor spaces, parks, gardens, and urban landscapes.

   e. Medicinal and Aromatic Plants: Studies the cultivation of plants with medicinal properties and aromatic compounds used in pharmaceuticals, cosmetics, and perfumery.

   f. Nursery Management: Focuses on the propagation, production, and distribution of seedlings, saplings, and ornamental plants.

   g. Post-Harvest Technology: Deals with techniques to extend the shelf life and quality of horticultural produce through proper handling, storage, and processing.

4. Economic Importance of Horticulture: Horticulture plays a crucial role in the global economy by contributing to food security, generating income for farmers, creating employment opportunities, and driving international trade. The cultivation and export of fruits, vegetables, and ornamental plants are significant contributors to the agricultural economy of many countries.

Conclusion: The definition and scope of horticulture reflect its diverse and significant contributions to society. From providing nutritious food to enhancing the beauty of landscapes, horticulture is an indispensable aspect of human life. As the world faces various challenges, such as climate change and population growth, the importance of horticulture in ensuring sustainable food production and environmental conservation becomes even more critical.

a. Nursery Management: Focuses on the propagation, production, and distribution of seedlings, saplings, and ornamental plants.

b. Post-Harvest Technology: Deals with techniques to extend the shelf life and quality of horticultural produce through proper handling, storage, and processing.

5. Economic Importance of Horticulture: Horticulture plays a crucial role in the global economy by contributing to food security, generating income for farmers, creating employment opportunities, and driving international trade. The cultivation and export of fruits, vegetables, and ornamental plants are significant contributors to the agricultural economy of many countries.

Conclusion: The definition and scope of horticulture reflect its diverse and significant contributions to society. From providing nutritious food to enhancing the beauty of landscapes, horticulture is an indispensable aspect of human life. As the world faces various challenges, such as climate change and population growth, the importance of horticulture in ensuring sustainable food production and environmental conservation becomes even more critical.

Chapter 2:

The Role of Horticulture in Early Civilizations

Introduction: Horticulture, the art and science of plant cultivation, played a pivotal role in the development of early civilizations. As ancient societies transitioned from nomadic lifestyles to settled communities, horticulture emerged as a fundamental practice that transformed human existence. This chapter explores the essential role of horticulture in shaping the growth, sustenance, and cultural fabric of early civilizations.

1. Transition to Settled Agriculture: The shift from hunter-gatherer societies to settled agriculture marked a profound turning point in human history. Horticulture, with its emphasis on intentional plant cultivation, allowed early humans to establish permanent settlements and move away from a nomadic lifestyle. This section delves into the key developments that led to the adoption of horticultural practices, including the domestication of wild plants and the emergence of agricultural techniques.

2. Foundations of Food Security: Horticulture provided a reliable and sustainable food source for early civilizations. This section explores the cultivation of staple crops, such as wheat, barley, rice, and maize, and their significance in ensuring food security. The development of agricultural systems, including irrigation and crop rotation, allowed civilizations to increase crop yields and support growing populations.

3. Cultural and Spiritual Significance: Beyond subsistence, horticulture held profound cultural and spiritual significance in early societies. Gardens and cultivated landscapes became symbols of abundance, prosperity, and human ingenuity. This section examines the role of horticulture in religious rituals, ceremonies, and artistic expressions, showcasing how gardens and cultivated spaces were integral to early civilizations' cultural identity.

4. Horticulture in Mesopotamia: The ancient civilizations of Mesopotamia, including Sumer, Babylon, and Assyria, relied heavily on horticulture for sustenance and economic prosperity. This section highlights the advancements in irrigation systems, the cultivation of date palms, and the establishment of lush gardens in the heart of Mesopotamian cities.

5. Egyptian Agricultural Marvels: Egyptian horticulture was an awe-inspiring example of early agricultural prowess. This section delves into the Nile River's significance in supporting agriculture and the cultivation of essential crops like wheat, barley, and flax. The iconic gardens of Egyptian temples and palaces, such as the gardens of Hatshepsut and the Hanging Gardens of Babylon, are explored.

6. Indus Valley's Agricultural Ingenuity: The Indus Valley Civilization, with its advanced urban planning and agricultural practices, provides a fascinating case study. This section highlights the Indus Valley's sophisticated drainage systems, agricultural terraces, and cultivation of diverse crops, reflecting their mastery of horticulture.

7. Chinese Agricultural Innovations: Ancient China's horticultural practices had a profound impact on world agriculture. This section delves into the Chinese invention of the seed drill, advancements in irrigation techniques, and the cultivation of rice, silk, tea, and other valuable crops.

The role of horticulture in early civilizations cannot be overstated. From providing sustenance and stability to shaping cultural identities and spiritual beliefs, horticulture laid the foundation for human civilization's development. The enduring legacy of early horticultural practices continues to inspire modern agricultural innovations and reminds us of the intimate connection between humanity and the earth.

Ancient Horticultural Practices in Mesopotamia, Egypt, China, and the Indus Valley

The ancient civilizations of Mesopotamia, Egypt, China, and the Indus Valley were pioneers in horticultural practices, shaping the development of agriculture and horticulture in their respective regions. This chapter explores the remarkable horticultural practices of these early societies, their agricultural innovations, and their profound impact on human civilization.

Mesopotamia: The Cradle of Horticulture

Introduction: Mesopotamia, often referred to as the "Cradle of Civilization," was a region located between the Tigris and Euphrates rivers in what is now modern-day Iraq. This fertile land and its unique geographical setting played a pivotal role in the development of agriculture and horticulture in ancient times. This chapter explores the significance of Mesopotamia's fertile plains, advancements in irrigation techniques, the cultivation of essential crops, and the establishment of remarkable gardens and orchards, including the legendary Hanging Gardens of Babylon.

1. The Fertile Plains of Mesopotamia: The fertile plains of Mesopotamia, nurtured by the annual flooding of the Tigris and Euphrates rivers, provided an abundant and dependable source of water and fertile soil for agriculture. This unique geographical setting allowed early civilizations to settle and develop agricultural practices, leading to the emergence of one of the world's earliest urban societies.
2. Advancements in Irrigation Techniques: Mesopotamian farmers made significant advancements in irrigation techniques to manage water resources effectively. They constructed intricate canal systems to distribute water to farmland, preventing waterlogging and ensuring optimal hydration for crops. Additionally, they built levees and dikes to protect against flooding and manage the water flow for irrigation purposes.
3. Cultivation of Essential Crops: Barley and wheat were the staple crops of Mesopotamia, forming the foundation of their diet and economy. Barley was widely used to produce bread and beer, an essential part of daily life. Wheat was also a staple crop, providing a valuable source of nutrition. In addition to grains, the Mesopotamians cultivated dates and various vegetables, showcasing their horticultural prowess.
4. Gardens and Orchards in Mesopotamia: Mesopotamian cities boasted magnificent gardens and orchards, showcasing the ingenuity and aesthetic sensibility of the civilization. The Hanging Gardens of Babylon, considered one of the Seven Wonders of the Ancient World, exemplified the splendor of their garden culture. While the existence of the Hanging Gardens is debated, historical texts and accounts suggest the presence of other splendid gardens and orchards throughout the region.

Mesopotamia's significance as the Cradle of Horticulture cannot be overstated. The fertile land between the Tigris and Euphrates rivers, coupled with advancements in irrigation, allowed for the cultivation of essential crops and the establishment of splendid gardens and orchards. The agricultural and horticultural practices of Mesopotamia laid the foundation for the development of subsequent civilizations, shaping the course of human history. The Hanging Gardens of Babylon, though shrouded in mystery, remain a symbol of the sophisticated horticultural achievements of this ancient civilization, leaving an enduring legacy that continues to inspire awe and wonder to this day.

Egypt: The Gift of the Nile

Egypt, an ancient civilization along the banks of the Nile River, owes its prosperity and cultural heritage to the Nile's life-sustaining waters. The Nile River, often referred to as "The Gift of the Nile," played a crucial role in supporting agriculture and horticulture in ancient Egypt. This chapter delves into the significance of the Nile River, the development of sophisticated irrigation systems, the cultivation of staple crops, and the profound importance of gardens in Egyptian culture, including their association with temple rituals and beliefs about the afterlife.

1. **The Nile River's Role in Supporting Agriculture and Horticulture**: The Nile River was the lifeblood of ancient Egypt, annually flooding its banks and depositing nutrient-rich silt, creating fertile lands known as the "Black Land." The inundation of the Nile provided a consistent water source and enabled farmers to cultivate crops even in arid regions. The predictable flooding pattern allowed for effective land management and the development of an agricultural calendar. The Nile's waters supported a flourishing agricultural society, making Egypt one of the most prosperous civilizations of its time.

2. **Development of Sophisticated Irrigation Systems**: To harness the Nile's waters, ancient Egyptians developed sophisticated irrigation systems. The shaduf, a counterweighted lever system, was used to lift water from the river and irrigate fields. Basin irrigation, an ingenious method, involved flooding fields through a network of canals and dikes, ensuring optimal water distribution. These irrigation techniques facilitated crop growth and allowed the cultivation of crops even in areas farther away from the Nile's banks.

3. **Cultivation of Staple Crops**: Wheat, barley, and flax were staple crops that formed the backbone of Egyptian agriculture. Wheat was a crucial food source and used to make bread, a dietary staple. Barley was also consumed, mainly as a source of food for animals. Flax was cultivated for its fiber, which was spun into linen, an essential textile used in clothing and other products. The cultivation of these crops was essential for the sustenance and economic prosperity of ancient Egypt.

4. **Importance of Gardens in Egyptian Culture**: Gardens held significant cultural and spiritual importance in ancient Egyptian society. Temple gardens were considered sacred spaces where religious rituals and offerings took place. They were carefully tended and filled with a variety of plants, symbolizing fertility, abundance, and rejuvenation. Gardens were also found in private spaces, providing aesthetic beauty and recreational areas for the elite. Additionally, gardens were a part of the Egyptian concept of the afterlife.

5. **Gardens and the Afterlife in Egyptian Beliefs**: In ancient Egyptian beliefs, gardens played a crucial role in the concept of the afterlife. The deceased were believed to transition to the "Field of Reeds" or the "Gardens of Aaru," where they would live an eternal and blissful existence. Burial sites often included gardens, reflecting the Egyptians' desire for prosperity in the afterlife. These "Gardens of Eternity" symbolized rejuvenation and eternal youth, with plants representing renewal and the cycle of life.

The Nile River was the lifeline of ancient Egypt, sustaining agriculture and horticulture and facilitating the growth of a flourishing civilization. The development of advanced irrigation systems allowed for efficient water management, while the cultivation of staple crops provided sustenance and economic stability. Gardens held a profound place in Egyptian culture, serving as sacred spaces, recreational areas, and symbols of the afterlife. The significance of the Nile and its impact on ancient Egypt's agricultural and horticultural practices continue to be an enduring legacy of one of the world's greatest civilizations.

China: Agricultural Marvels of the East

Introduction: Ancient China, with its diverse geography and rich cultural heritage, witnessed remarkable advancements in agriculture and horticulture. This chapter explores the influence of China's diverse geography on agricultural practices, innovations in agricultural tools and techniques, the cultivation of rice as a transformative staple crop, and the significance of tea, silk, and other horticultural products in Chinese culture.

1. **The Diverse Geography of Ancient China and Its Influence on Agriculture**: China's geography varies from fertile plains to mountainous regions, providing a diverse range of ecosystems for agriculture. The fertile floodplains of major rivers, such as the Yellow River and Yangtze River, facilitated the development of agrarian civilizations. The mountainous areas required innovative agricultural practices to maximize land use. The Chinese adapted their farming techniques to suit the specific demands of each region, making agriculture a foundation of Chinese civilization.

2. **Innovations in Agricultural Tools and Techniques**: Ancient Chinese farmers developed innovative tools and techniques to enhance agricultural productivity. The seed drill, credited to the Chinese agriculturalist Jia Sixie, allowed for precise seed placement, resulting in improved crop yields. The use of iron ploughs revolutionized soil preparation, making farming more efficient and reducing labor intensity. These inventions played a crucial role in China's agricultural marvels.

3. **The Cultivation of Rice:** Transforming China's Agricultural Landscape: Rice cultivation, especially in the Yangtze River Basin and southern regions of China, played a transformative role in Chinese agriculture. The introduction of wet-rice farming techniques allowed for multiple crop cycles and high yields. The surplus production of rice contributed to population growth, urbanization, and the flourishing of ancient Chinese civilizations.

4. **The Significance of Tea, Silk, and Other Horticultural Products in Chinese Culture**: Tea, silk, and other horticultural products held immense cultural and economic importance in ancient China. Tea cultivation became an integral part of Chinese culture, with tea being consumed for its medicinal properties and social significance. Silk production from silkworms became a coveted industry, contributing to China's wealth and cultural identity. Additionally, horticultural products such as oranges, peaches, and bamboo held symbolic meanings in Chinese traditions.

China's agricultural marvels in ancient times were shaped by its diverse geography, innovative farming techniques, and the cultivation of transformative crops like rice. The Chinese horticultural products, such as tea and silk, held profound cultural significance and contributed to the nation's wealth and identity. The agricultural achievements of ancient China laid the foundation for its subsequent agrarian-based civilizations and continue to influence modern agricultural practices. China's historical contributions to agriculture and horticulture stand as a testament to human ingenuity and the close relationship between humans and the land they cultivate.

Indus Valley: Urban Planning and Agricultural Prowess

Introduction: The Indus Valley Civilization, one of the ancient world's earliest urban societies, thrived in the fertile plains of the Indus River basin in present-day Pakistan and northwest India.

This chapter explores the unique urban planning of the Indus Valley, its impact on agriculture, advances in agricultural techniques, the cultivation of staple crops like wheat, barley, rice, and cotton, as well as the cultural and spiritual significance of gardens and the use of plants in religious rituals.

1. **The Urban Planning of the Indus Valley Civilization and Its Impact on Agriculture**: The Indus Valley Civilization exhibited remarkable urban planning with well-organized cities featuring an efficient drainage and sanitation system. The planned layout of cities, including Mohenjo-Daro and Harappa, facilitated the administration of agricultural lands and ensured a steady food supply to support the urban population. The systematic division of land into grids and well-designed streets enabled effective land management and equitable distribution of agricultural resources.

2. **Advances in Agricultural Techniques**: Terraced Farming and Flood Control: The Indus Valley Civilization demonstrated ingenious agricultural techniques to maximize productivity in the fertile but flood-prone Indus River basin. Terraced farming on hilly terrains allowed for cultivation in regions with limited arable land, preventing soil erosion and conserving water. Additionally, the construction of flood control systems, such as embankments and reservoirs, mitigated the adverse impact of seasonal floods, ensuring sustainable agriculture.

3. **Cultivation of Wheat, Barley, Rice, and Cotton in the Indus Valley**: The Indus Valley people practiced agriculture on an extensive scale, growing a variety of staple crops. Wheat and barley were primary cereal crops, forming the dietary backbone of the civilization. Rice cultivation flourished in the lower Indus region, benefiting from the fertile alluvial soil and the Indus River's abundant water supply. Cotton was another essential crop, contributing to the thriving textile industry.

4. **Cultural and Spiritual Significance of Gardens and Plants in Religious Rituals**: Gardens held cultural and spiritual significance in the Indus Valley Civilization. Terraced gardens adorned with ornamental plants and trees reflected the people's appreciation for aesthetics and nature's beauty. Gardens were also associated with religious rituals, and certain plants played crucial roles in religious practices. The "Pipal" or sacred fig tree was venerated, and the "Swastika" symbol held spiritual significance, being associated with various religious beliefs.

The Indus Valley Civilization stands as a testament to the early human's mastery of urban planning and agricultural prowess. The well-organized cities and advanced agricultural techniques enabled this ancient civilization to thrive sustainably in the fertile Indus River basin. The cultivation of staple crops like wheat, barley, rice, and cotton, along with the cultural significance of gardens and plants in religious rituals, reveals the deep connection between the Indus Valley people and the land they cultivated. The legacy of the Indus Valley Civilization continues to inspire modern agricultural practices, highlighting the enduring impact of early agricultural advancements on human civilization.

Importance and Role of Horticulture in Society in Developing Countries

Introduction: Horticulture, the science and art of cultivating fruits, vegetables, flowers, and ornamental plants, plays a vital role in developing countries. This chapter explores the importance and multifaceted role of horticulture in society, focusing on how it contributes to food security, economic growth, environmental sustainability, employment generation, and cultural enrichment in developing nations.

1. **Food Security and Nutrition**: Horticulture significantly contributes to food security in developing countries by diversifying the diet and providing essential vitamins, minerals, and nutrients. Fruits and vegetables cultivated in horticultural practices offer a rich source of micronutrients, promoting better nutrition and combating malnutrition. Small-scale horticultural production, particularly in peri-urban areas, ensures access to fresh and perishable produce, enhancing food security and reducing dependence on imported foods.

2. **Economic Growth and Livelihoods**: Horticulture serves as a powerful economic driver in developing countries. It provides employment opportunities along the entire value chain, from production to processing, packaging, distribution, and marketing. Smallholder farmers, especially women and rural communities, benefit from horticulture, as it offers income-generating opportunities and fosters entrepreneurship. The export of horticultural products also boosts foreign exchange earnings and contributes to the overall economic growth of the country.

3. **Environmental Sustainability and Biodiversity**: Horticulture plays a crucial role in promoting environmental sustainability in developing nations. Diverse cropping patterns in horticultural practices encourage ecological balance, reducing the risk of soil degradation and pest outbreaks. Agroforestry and fruit orchards enhance biodiversity by creating habitats for various plant andanimal species. Additionally, sustainable horticultural practices, such as organic farming, conserve soil and water resources, mitigating the environmental impact of agriculture.

4. **Adaptation to Climate Change**: In the face of climate change, horticulture offers adaptive strategies for developing countries. Short gestation periods of horticultural crops allow farmers to respond quickly to changing climatic conditions. Climate-resilient horticultural varieties can withstand extreme weather events, enhancing farmers' resilience to climatic uncertainties and reducing crop losses.

5. **Cultural Enrichment and Tourism**: Horticulture adds to the cultural richness of developing societies. Traditional practices, such as flower festivals and garden showcases, promote cultural heritage and celebrate local customs. Beautifully landscaped gardens and parks attract tourists, contributing to the growth of the tourism industry and generating revenue for the local economy.

6. **Empowerment of Women**: Horticulture often provides opportunities for women's empowerment in developing countries. Women play a significant role in horticultural activities, from planting and nurturing crops to marketing produce. By engaging in horticulture, women gain economic independence, social recognition, and decision-making power within their households and communities.

The importance and role of horticulture in society in developing countries cannot be overstated. It serves as a pillar of food security, economic growth, and environmental sustainability, while also enriching cultural heritage and empowering marginalized communities. By recognizing and investing in horticultural development, developing nations can unlock tremendous potential for prosperity, food sovereignty, and a sustainable future for generations to come.

Chapter 3:

Branches of Horticulture

Introduction: Horticulture, a diverse and multidisciplinary field, encompasses various sub-disciplines that focus on specific aspects of plant cultivation, management, and utilization. This chapter delves into the key sub-disciplines of horticulture, exploring their unique contributions to agricultural practices, environmental conservation, and societal well-being.

1. Pomology: Pomology is the branch of horticulture dedicated to the study of fruit crops. It involves the cultivation, breeding, and management of fruit-bearing plants such as apples, pears, citrus fruits, and berries. Pomologists work to improve fruit quality, enhance resistance to diseases and pests, and develop new fruit varieties with desirable traits. This sub-discipline plays a crucial role in ensuring a bountiful supply of nutritious fruits and contributing to the fruit industry's economic growth.

2. Olericulture: Olericulture focuses on the science of vegetable production and management. It involves the cultivation of vegetables such as tomatoes, cucumbers, lettuce, and carrots. Olericulturists work to optimize vegetable growth, improve post-harvest handling, and enhance nutritional content. This sub-discipline is essential for sustaining the availability of diverse and nutrient-rich vegetables for human consumption.

3. Floriculture: Floriculture is concerned with the cultivation and management of ornamental plants, particularly flowers. Floriculturists grow and breed flowering plants like roses, lilies, chrysanthemums, and orchids. This sub-discipline contributes to the global floral industry, encompassing cut flowers, potted plants, and landscaping elements. Floriculture enhances the aesthetic appeal of landscapes and plays a significant role in cultural events, celebrations, and rituals.

4. Landscape Horticulture: Landscape horticulture focuses on the design, establishment, and maintenance of outdoor spaces, including gardens, parks, and urban landscapes. Landscape horticulturists combine artistic elements with ecological principles to create visually appealing and functional landscapes. This sub-discipline contributes to urban green spaces, biodiversity conservation, and environmental restoration efforts.

5. Turfgrass Management: Turfgrass management involves the cultivation and care of grasses used in sports fields, golf courses, lawns, and recreational areas. Turfgrass managers optimize grass growth, ensure uniformity, and address challenges related to pest and disease control. This sub-discipline plays a critical role in providing safe and high-quality playing surfaces while minimizing environmental impact.

6. Medicinal and Aromatic Plants: The sub-discipline of medicinal and aromatic plants focuses on the cultivation and utilization of plants with medicinal and aromatic properties. Horticulturists in this field work to produce herbs and aromatic plants, such as lavender, peppermint, and chamomile, which have therapeutic and aromatic uses in pharmaceuticals, cosmetics, and traditional medicine.

7. Post-harvest management is a significant sub-discipline of horticulture that focuses on the handling, preservation, and storage of horticultural crops after they have been harvested from the field. This discipline plays a crucial role in ensuring the quality and marketability of horticultural produce, minimizing losses, and extending the shelf life of perishable products. Post-harvest

management involves a series of practices and technologies aimed at maintaining the freshness, nutritional content, and overall quality of fruits, vegetables, and other horticultural commodities from the time of harvest to the point of consumption or processing.

The diverse sub-disciplines of horticulture collectively contribute to sustainable agriculture, environmental conservation, and human well-being. Pomology, olericulture, floriculture, landscape horticulture, turfgrass management, and medicinal and aromatic plants address specific aspects of plant cultivation and utilization, ensuring a secure food supply, enhancing aesthetic landscapes, and promoting health and well-being. These sub-disciplines highlight the significance of horticulture in meeting diverse societal needs and fostering a sustainable and prosperous future.

Chapter 4

Plant Morphology and Anatomy

Introduction: Plant morphology and anatomy are fundamental branches of plant biology that explore the structure and growth of plants at the cellular, tissue, and whole-plant levels. This chapter provides an in-depth understanding of plant morphology, encompassing the diverse forms and structures that plants exhibit, and plant anatomy, which delves into the internal organization of plant tissues and cells.

1. **Plant Structure**: The structure of a plant encompasses various organs and tissues that collectively contribute to its growth and function. The main plant organs include roots, stems, leaves, flowers, and fruits. Roots anchor the plant in the soil, absorb water and nutrients, and store reserve substances. Stems provide support and transport water, nutrients, and photosynthetic products throughout the plant. Leaves are the primary sites for photosynthesis, converting sunlight into energy, and flowers are reproductive structures responsible for seed production. Fruits are the mature ovary of a flower, protecting seeds and facilitating seed dispersal.

2. **Plant Growth**: Plant growth is a dynamic process that involves cell division, cell elongation, and cell differentiation. Meristems, regions of active cell division, are found at the tips of roots and shoots, enabling primary growth. Apical meristems are responsible for the growth in length, while lateral meristems contribute to growth in girth. Secondary growth occurs in woody plants due to the activity of vascular cambium and cork cambium, adding thickness to stems and roots. The regulation of growth hormones, such as auxins, gibberellins, and cytokinins, influences plant growth patterns.

3. **Plant Tissues**: Plant tissues are groups of cells with similar functions and characteristics. There are three main types of plant tissues: dermal, ground, and vascular. Dermal tissue, found on the outer surfaces of plant organs, forms the epidermis and cuticle, providing protection and controlling water loss. Ground tissue makes up the bulk of plant tissue and performs functions like photosynthesis, storage, and support. Vascular tissue consists of xylem and phloem, responsible for water and nutrient transport throughout the plant.

4. **Leaf Morphology**: Leaves are essential photosynthetic organs with diverse morphologies. Their shape, size, and arrangement vary among plant species. Leaf venation patterns, such as parallel and reticulate, also differ and aid in species identification. Different leaf adaptations, like needle-shaped leaves in conifers or succulent leaves in desert plants, reflect their ecological roles and adaptations to environmental conditions.

5. **Stem Anatomy**: Stems exhibit distinct anatomical features, with variations depending on the plant type. Herbaceous stems consist of primary tissues and lack secondary growth, whereas woody stems have secondary growth and additional tissues like wood and bark. The arrangement of vascular bundles, pith, cortex, and cambium layers contribute to stem anatomy and function.

6. **Flower Anatomy**: Flowers are reproductive structures responsible for seed production and the continuation of plant generations. A typical flower consists of four main whorls: sepals, petals, stamens, and carpels. Sepals protect the developing bud, while petals attract pollinators. Stamens produce pollen containing male gametes, and carpels house the ovules where female gametes

develop. Flowers exhibit great diversity in size, shape, and color, depending on plant species and pollination strategies.

7. **Fruit Anatomy**: Fruits are mature ovaries that develop from fertilized flowers and protect seeds. Fruit anatomy includes the pericarp, comprising three layers: the exocarp, mesocarp, and endocarp. The pericarp protects seeds and aids in seed dispersal. Different types of fruits, such as fleshy fruits, dry fruits, and aggregate fruits, showcase a wide array of adaptations for seed dispersal and survival.

Plant morphology and anatomy offer insights into the intricate structures and growth processes of plants. Understanding plant structure and growth is essential for horticulture, agriculture, and ecological research, as it influences how plants respond to environmental conditions, nutrient uptake, and overall plant health. The comprehensive knowledge of plant morphology and anatomy enhances our appreciation of the diversity and complexity of the plant kingdom and underpins advancements in agriculture, horticulture, and conservation efforts.

Types of Horticultural Crops Based on Their Bearing Habits

Horticultural crops exhibit various bearing habits, referring to their patterns of fruiting or flowering throughout the year. These habits play a crucial role in determining the crop's production cycle, harvest times, and overall management practices. Understanding the different types of horticultural crops based on their bearing habits helps horticulturists, farmers, and gardeners plan and optimize their crop production. In this chapter, we will explore the main types of horticultural crops based on their bearing habits.

Annual Crops:

Annual crops complete their life cycle within a single growing season. They are sown, grow, flower, and produce fruits or seeds, all in the same year. After harvesting, the plants die, and new plants are grown from seeds in the subsequent season. Common examples of annual horticultural crops include:

1. Tomato (*Solanum lycopersicum*)

2. Cucumber (*Cucumis sativus*)

3. Spinach (*Spinacia oleracea*)

4. Sunflower (*Helianthus annuus*)

5. Radish (*Raphanus sativus*)

Biennial Crops:

Biennial crops have a two-year life cycle. They typically grow vegetatively in the first year, forming roots, leaves, and stems. In the second year, they flower, set seeds, and die. Biennial crops often require vernalization, a period of cold temperatures, to induce flowering. Examples of biennial horticultural crops include:

1. Carrot (*Daucus carota*)

2. Beetroot (*Beta vulgaris*)

3. Parsley (*Petroselinum crispum*)

4. Cabbage (*Brassica oleracea var. capitata)*

Perennial Crops:

Perennial crops have a life cycle that extends for multiple years, allowing them to flower and fruit for several seasons. Once established, perennial crops continue to grow and produce, making them valuable for long-term horticultural purposes. Common examples of perennial horticultural crops include:

1. Apple (*Malus domestica*)

2. Pear (*Pyrus communis*)

3. Grapevine (*Vitis vinifera*)

4. Strawberry (*Fragaria* × *ananassa*)

5. Blueberry (*Vaccinium corymbosum*)

Seasonal Crops:

Seasonal crops exhibit a specific bearing habit during certain times of the year, but not throughout the entire year. They may be annual, biennial, or perennial, but their flowering or fruiting periods are limited to specific seasons. Seasonal horticultural crops can be further categorized based on the seasons they bear:

Summer Season Crops:

- Watermelon (*Citrullus lanatus*)

- Okra (*Abelmoschus esculentus*)

- Eggplant (*Solanum melongena*)

Monsoon Season Crops:

- Rice (*Oryza sativa*)

- Turmeric (*Curcuma longa*)

- Ginger (*Zingiber officinale*)

Autumn/Fall Season Crops:

- Pumpkin (*Cucurbita pepo*)
- Apple (*Malus domestica*)
- Pomegranate (*Punica granatum*)

Winter Season Crops:

- Cauliflower (*Brassica oleracea* var. botrytis)
- Cabbage (*Brassica oleracea* var. *capitata*)
- Carrot (*Daucus carota*)

Spring Season Crops:

- Lettuce (*Lactuca sativa*)
- Spinach (*Spinacia oleracea*)
- Strawberry (*Fragaria* × *ananassa*)

Understanding the different types of horticultural crops based on their bearing habits is crucial for effective crop planning, rotation, and management. By selecting a diverse range of crops with varying bearing habits, horticulturists can maintain a continuous supply of fresh produce throughout the year and optimize the utilization of their growing spaces. Additionally, knowledge of the bearing habits helps in proper timing for pruning, fertilization, and pest management, ensuring healthy and productive crops.

Fruit crops are a diverse group of plants cultivated for their delicious and nutritious fruits. They are an essential part of human diets and contribute significantly to the economy and agricultural industry. Fruit crops encompass a wide range of flavors, textures, and colors, delighting taste buds and providing essential vitamins and minerals. In this chapter, we will explore different types of fruit crops, including their common names, scientific names, and families.

Citrus Fruits:

1. Orange (Common Name): *Citrus sinensis* (Scientific Name) - Rutaceae (Family)

2. Lemon (Common Name): *Citrus limon* (Scientific Name) - Rutaceae (Family)

3. Lime (Common Name): *Citrus aurantiifolia* (Scientific Name) - Rutaceae (Family)

4. Grapefruit (Common Name): *Citrus paradisi* (Scientific Name) - Rutaceae (Family)

5. Mandarin (Common Name): *Citrus reticulata* (Scientific Name) - Rutaceae (Family)

Berry Fruits:

1. Strawberry (Common Name): *Fragaria × ananassa* (Scientific Name) - Rosaceae (Family)
2. Blueberry (Common Name): *Vaccinium corymbosum* (Scientific Name) - Ericaceae (Family)
3. Raspberry (Common Name): *Rubus idaeus* (Scientific Name) - Rosaceae (Family)
4. Blackberry (Common Name): *Rubus fruticosus* (Scientific Name) - Rosaceae (Family)
5. Cranberry (Common Name): *Vaccinium macrocarpon* (Scientific Name) - Ericaceae (Family)

Stone Fruits:

1. Peach (Common Name): *Prunus persica* (Scientific Name) - Rosaceae (Family)
2. Plum (Common Name): *Prunus domestica* (Scientific Name) - Rosaceae (Family)
3. Apricot (Common Name): *Prunus armeniaca* (Scientific Name) - Rosaceae (Family)
4. Cherry (Common Name): *Prunus avium* (Scientific Name) - Rosaceae (Family)
5. Nectarine (Common Name): *Prunus persica* var. *nucipersica* (Scientific Name) - Rosaceae (Family)

Tropical Fruits:

1. Mango (Common Name): *Mangifera indica* (Scientific Name) - Anacardiaceae (Family)
2. Banana (Common Name): *Musa acuminata* (Scientific Name) - Musaceae (Family)
3. Pineapple (Common Name): *Ananas comosus* (Scientific Name) - Bromeliaceae (Family)
4. Papaya (Common Name): *Carica papaya* (Scientific Name) - Caricaceae (Family)
5. Avocado (Common Name): *Persea americana* (Scientific Name) - Lauraceae (Family)

Pome Fruits:

1. Apple (Common Name): *Malus domestica* (Scientific Name) - Rosaceae (Family)
2. Pear (Common Name): *Pyrus communis* (Scientific Name) - Rosaceae (Family)
3. Quince (Common Name): *Cydonia oblonga* (Scientific Name) - Rosaceae (Family)

Melon and Gourd Fruits:

1. Watermelon (Common Name): *Citrullus lanatus* (Scientific Name) - Cucurbitaceae (Family)
2. Cantaloupe (Common Name): *Cucumis melo* (Scientific Name) - Cucurbitaceae (Family)
3. Pumpkin (Common Name): *Cucurbita pepo* (Scientific Name) - Cucurbitaceae (Family)
4. Squash (Common Name): *Cucurbita* spp. (Scientific Name) - Cucurbitaceae (Family)

Tropical and Subtropical Fruits:

1. Guava (Common Name): *Psidium guajava* (Scientific Name) - Myrtaceae (Family)
2. Kiwi (Common Name): *Actinidia deliciosa* (Scientific Name) - Actinidiaceae (Family)
3. Lychee (Common Name): *Litchi chinensis* (Scientific Name) - Sapindaceae (Family)
4. Dragon Fruit (Common Name): *Hylocereus undatus* (Scientific Name) - Cactaceae (Family)

Fruit crops are not only a source of nutrition and delight but also play a significant role in agricultural and economic development. By understanding the different types of fruit crops, their scientific classifications, and their families, horticulturists and researchers can better appreciate the diversity and importance of these plants. The cultivation and sustainable management of fruit crops contribute to food security, promote biodiversity, and enrich our diets with a wide array of flavors and nutritional benefits.

Flower Crops: Types, Common Names, Scientific Names, and Families

Flower crops are a diverse group of plants cultivated primarily for their ornamental beauty and aesthetic appeal. These crops brighten gardens, landscapes, and floral arrangements, adding colors, scents, and visual interest. Flower crops are integral to the floriculture industry, contributing to celebrations, events, and expressions of emotions. In this chapter, we will explore different types of flower crops, including their common names, scientific names, and families.

Rose (Common Name): Rosa spp. (Scientific Name) Rose Family (Rosaceae):

Daisy (Common Name): *Bellis perennis* (Scientific Name)
Sunflower (Common Name): *Helianthus annuus* (Scientific Name)
Chrysanthemum (Common Name): *Chrysanthemum* spp. (Scientific Name)
Aster (Common Name): *Aster* spp. (Scientific Name)

Orchid (Common Name): Orchidaceae spp. (Scientific Name) Family (Orchidaceae):
Snapdragon (Common Name): Antirrhinum majus (Scientific Name) Family (Plantaginaceae)
Carnation (Common Name): Dianthus caryophyllus (Scientific Name) Family (Caryophyllaceae)
Iris (Common Name): Iris spp. (Scientific Name) (Iridaceae) Family

(Amaryllidaceae), Daffodil (Common Name): Narcissus spp. (Scientific Name) Family
Hyacinth (Common Name): *Hyacinthus orientalis* (Scientific Name) (Asparagaceae) Family
Flower crops bring joy and beauty to our lives, whether adorning gardens, brightening bouquets, or gracing special occasions. Understanding the common names, scientific names, and families of these flower crops enhances our appreciation of their diverse forms and characteristics. The floriculture industry relies on the cultivation and propagation of these ornamental plants, contributing to the cultural and economic significance of flowers worldwide. Horticulturists, floral designers, and flower enthusiasts can celebrate the unique beauty and variety of flower crops as they continue to nurture and cherish these beloved plants.

Vegetable Crops: Types, Common Names, Scientific Names, and Families

Vegetable crops are an essential part of our daily diet, providing us with essential vitamins, minerals, and nutrients. These crops encompass a wide variety of plants that are cultivated for their edible parts, including roots, stems, leaves, flowers, and fruits. Vegetable crops contribute to food security, culinary diversity, and healthy nutrition. In this chapter, we will explore different types of vegetable crops, including their common names, scientific names, and families.
Brassicaceae Family:

1. Cabbage (Common Name): *Brassica oleracea* var. capitata (Scientific Name)

2. Broccoli (Common Name): *Brassica oleracea* var. italica (Scientific Name)

3. Cauliflower (Common Name): *Brassica oleracea* var. botrytis (Scientific Name)

4. Brussels Sprouts (Common Name): *Brassica oleracea* var. gemmifera (Scientific Name)

5. Kale (Common Name): *Brassica oleracea* var. *sabellica* (Scientific Name)
Solanaceae Family:
1. Tomato (Common Name): *Solanum lycopersicum* (Scientific Name)

2. Potato (Common Name): *Solanum tuberosum* (Scientific Name)

3. Eggplant (Common Name): *Solanum melongena* (Scientific Name)

4. Bell Pepper (Common Name): *Capsicum annuum* (Scientific Name)

Cucurbitaceae Family:

1. Cucumber (Common Name): *Cucumis sativus* (Scientific Name)

2. Pumpkin (Common Name): *Cucurbita pepo* (Scientific Name)

3. Squash (Common Name): *Cucurbita spp.* (Scientific Name)

4. Watermelon (Common Name): *Citrullus lanatus* (Scientific Name)

Apiaceae Family:

1. Carrot (Common Name): *Daucus carota* (Scientific Name)

2. Celery (Common Name): *Apium graveolens* (Scientific Name)

3. Parsley (Common Name): *Petroselinum crispum* (Scientific Name)

Fabaceae Family:

1. Peas (Common Name): *Pisum sativum* (Scientific Name)

2. Beans (Common Name): *Phaseolus vulgaris* (Scientific Name)

3. **Lentils** (Common Name): *Lens culinaris* (Scientific
   Name) Asteraceae Family:
1. **Lettuce** (Common Name): *Lactuca sativa* (Scientific Name)
   Alliaceae Family:
1. Onion (Common Name): *Allium cepa* (Scientific Name)

2. Garlic (Common Name): *Allium sativum* (Scientific Name)
   Amaranthaceae Family:
1. Spinach (Common Name): *Spinacia oleracea* (Scientific Name)
   Chenopodiaceae Family:
1. Beetroot (Common Name): *Beta vulgaris* (Scientific Name)
   Apiaceae Family:
1. Radish (Common Name): *Raphanus sativus* (Scientific Name)

Vegetable crops form the foundation of healthy and balanced diets, providing us with an array of tastes, textures, and culinary possibilities. Understanding the common names, scientific names, and families of these vegetable crops helps us appreciate the diversity of the plant kingdom and the role that these crops play in sustaining human life. The cultivation and consumption of vegetable crops are essential for ensuring food security, promoting sustainable agriculture, and fostering a healthier and more nutrititive society.

Chapter 5

Plant Physiology

Introduction: Plant physiology is a branch of botany that deals with the study of various physiological processes and functions in plants. This chapter delves into plant water relations, a fundamental aspect of plant physiology, and explores the process of transpiration, which plays a vital role in water movement and plant survival.

Plant Water Relations and Transpiration

1. **Water Uptake and Transport in Plants**: Water is essential for plant growth and development, and plants absorb water from the soil through their roots. The process of water uptake is facilitated by the root hairs, which increase the surface area for water absorption. Once inside the plant, water is transported upwards through the xylem, a complex tissue specialized for water conduction.
2. **Cohesion-Tension Theory**: The cohesion-tension theory explains how water moves upward through the xylem vessels. According to this theory, water molecules form a continuous column due to cohesion, with hydrogen bonding between adjacent water molecules. Additionally, the transpiration of water from the leaves creates a tension that pulls water upwards, leading to a continuous flow from the roots to the shoots.
3. **Transpiration**: Transpiration is the process by which water vapor escapes from the aerial parts of the plant, primarily through small pores called stomata on the leaf surface. This loss of water vapor is a crucial part of the plant's cooling mechanism and helps maintain cell turgidity. Transpiration is directly related to environmental factors such as temperature, humidity, wind speed, and light intensity.
4. **Factors Affecting Transpiration Rate**: Several factors influence the rate of transpiration in plants. Higher temperatures and low humidity can increase transpiration as they create a steeper water vapor concentration gradient between the leaf surface and the atmosphere. Wind enhancestranspiration by removing water vapor from the leaf surface, while light intensity affects stomatal opening, influencing the rate of water loss.
5. **Stomatal Regulation**: Stomata are specialized pores found in the epidermis of leaves and stems, responsible for gas exchange and water vapor release. Stomatal opening and closure are regulated by various factors, including light, temperature, humidity, and internal plant signals such as water availability and hormone levels. This regulation ensures that transpiration rates are balanced to prevent excessive water loss and maintain proper water status in the plant.
6. **Significance of Transpiration**: Transpiration plays a crucial role in plant physiology and overall plant health. It facilitates nutrient uptake from the soil, maintains cell turgidity, cools the plant, and enables the movement of water and nutrients from the roots to the shoots. Transpiration also creates a negative pressure gradient within the xylem, supporting water movement against gravity.

Plant water relations and transpiration are fundamental processes that govern water movement and plant survival. Understanding these processes is essential for horticulturists, farmers, and researchers to manage water resources effectively, optimize irrigation practices, and enhance crop productivity. Proper management of plant water relations is critical for sustainable agriculture and the maintenance of healthy and thriving plant communities in various ecological settings.

Photosynthesis and Respiration

Photosynthesis and respiration are two fundamental processes that govern energy exchange in plants. This chapter delves into the intricate mechanisms of these processes, exploring how plants capture solar energy, convert it into chemical energy, and utilize it to sustain life and growth.

1. **Photosynthesis**: Photosynthesis is the process by which plants, algae, and some bacteria convert sunlight, water, and carbon dioxide into glucose (a sugar) and oxygen. It takes place in chloroplasts, specialized organelles found in plant cells. Chlorophyll, a green pigment within chloroplasts, captures sunlight, initiating the series of chemical reactions that drive photosynthesis.

2. **Light Reactions**: Photosynthesis begins with the light reactions, which occur in the thylakoid membranes of chloroplasts. During this phase, light energy is absorbed by chlorophyll, leading to the splitting of water molecules and the release of oxygen. Additionally, this process generates high-energy molecules such as ATP (adenosine triphosphate) and NADPH (nicotinamide adenine dinucleotide phosphate), which serve as energy carriers for the next phase.

3. **Calvin Cycle (Dark Reactions)**: The Calvin Cycle, also known as the dark reactions, occurs in the stroma of chloroplasts. It involves the fixation of carbon dioxide and the subsequent reduction of carbon compounds using ATP and NADPH generated during the light reactions. The end product of the Calvin Cycle is glucose, which is crucial for plant growth and sustenance.

4. **Factors Affecting Photosynthesis**: Several factors influence the rate of photosynthesis, including light intensity, carbon dioxide concentration, temperature, and the availability of water and nutrients. Optimal levels of these factors enhance photosynthesis, leading to increased plant growth and productivity.

5. **Respiration**: Respiration is the process by which plants (and other living organisms) break down glucose and other organic molecules to release energy for cellular activities. Respiration occurs in the mitochondria, where glucose is oxidized in the presence of oxygen to produce carbon dioxide, water, and ATP, which serves as the primary energy currency of cells.

6. **Aerobic Respiration**: Aerobic respiration is the most common form of respiration in plants and occurs in the presence of oxygen. During aerobic respiration, glucose is completely oxidized, producing a large amount of ATP for cellular processes.

7. **Anaerobic Respiration**: In some instances, plants undergo anaerobic respiration when oxygen is limited or absent. Anaerobic respiration produces less ATP and results in the accumulation of metabolic by-products, such as ethanol or lactic acid.

8. **Balance between Photosynthesis and Respiration**: Photosynthesis and respiration are interconnected processes in plants. During the day, when sunlight is available, photosynthesis predominates, leading to the production of glucose and oxygen. At night or in low-light conditions, photosynthesis slows down, and respiration becomes more significant in utilizing stored energy.

Photosynthesis and respiration are fundamental processes that sustain life and energy balance in plants. Understanding the intricacies of these processes is crucial for horticulturists, farmers, and researchers to optimize crop growth, improve agricultural practices, and ensure the efficient utilization of resources. The balance between photosynthesis and respiration influences plant growth, productivity, and overall health, making these processes essential considerations in horticultural practices and plant biology research.

Plant Nutrition and Nutrient Uptake

Introduction: Plant nutrition is a critical aspect of horticulture that focuses on understanding how plants obtain and utilize essential nutrients for growth and development. This chapter explores the diverse range of nutrients required by plants, the processes involved in nutrient uptake, and the factors that influence nutrient availability in the soil.

1. **Essential Nutrients for Plants**: Plants require a variety of nutrients for their metabolic processes, growth, and reproduction. These essential nutrients can be broadly categorized into two groups: macronutrients and micronutrients. Macronutrients, such as nitrogen, phosphorus, potassium, calcium, magnesium, and sulfur, are required in relatively large quantities. Micronutrients, including iron, manganese, zinc, copper, boron, molybdenum, and chlorine, are needed in smaller amounts but are equally crucial for plant health.
2. **Nutrient Sources for Plants**: Plants acquire nutrients from various sources, primarily from the soil. Nutrients are present in the soil solution, bound to soil particles, or in organic matter. Root hairs and mycorrhizal associations play essential roles in nutrient uptake. In addition to soil, some plants can obtain nutrients from the atmosphere through foliar absorption.
3. **Nutrient Uptake Mechanisms**: The process of nutrient uptake involves several mechanisms, including active and passive uptake. Active uptake relies on the energy derived from ATP hydrolysis to transport nutrients against concentration gradients into plant cells. Passive uptake, on the other hand, occurs when nutrients move into plant cells along concentration gradients without the need for energy expenditure.
4. **Root Architecture and Nutrient Acquisition**: The efficiency of nutrient uptake is influenced by root architecture. Plants with extensive root systems are better equipped to explore a larger volume of soil, increasing the likelihood of encountering nutrient-rich areas. Fine root hairs increase the surface area for nutrient absorption.
5. **Nutrient Availability in Soil**: The availability of nutrients in the soil is influenced by various factors, including soil pH, temperature, soil moisture, organic matter content, and the presence of other ions that can compete with plant nutrients for uptake. Soil management practices can significantly impact nutrient availability and uptake efficiency.
6. **Nutrient Deficiency and Toxicity**: Imbalances in nutrient availability can lead to nutrient deficiencies or toxicities in plants. Nutrient deficiencies manifest as characteristic symptoms, such as chlorosis (yellowing of leaves), necrosis (death of plant tissues), or stunted growth. Excessive nutrient uptake can result in nutrient toxicity, leading to physiological and structural abnormalities.
7. **Fertilization Practices**: To optimize plant nutrition, horticulturists and farmers use fertilizers to supplement nutrient requirements. Fertilizers are formulated to provide a balanced combination of essential nutrients, and their application must be carefully managed to avoid overuse and environmental pollution.

Plant nutrition and nutrient uptake is essential for sustainable and productive horticulture. Proper nutrient management ensures optimal plant growth, improved crop yields, and healthier plants.

By considering nutrient requirements, soil conditions, and fertilization practices, horticulturists can foster robust plant health and contribute to environmentally responsible agricultural practices.

Hormonal Regulation and Plant Growth

Introduction: Hormonal regulation plays a crucial role in controlling various aspects of plant growth and development. This chapter explores the diverse group of plant hormones, their functions, and the intricate ways in which they interact to govern different physiological processes in plants.

1. **Plant Hormones**: Plant hormones, also known as phytohormones, are chemical messengers produced in specific plant tissues. The major classes of plant hormones include auxins, gibberellins, cytokinins, abscisic acid, ethylene, and brassinosteroids. Each hormone has distinct roles in regulating various aspects of plant growth and development.
2. **Auxins**: Auxins are primarily produced in the apical meristems and young leaves. They play a central role in cell elongation, promoting stem and root growth. Auxins also regulate apical dominance, promoting the growth of the main shoot while inhibiting lateral bud development.
3. **Gibberellins**: Gibberellins are synthesized in growing plant organs, such as young leaves and developing seeds. They promote stem elongation, fruit growth, and seed germination. Gibberellins are essential for breaking seed dormancy and inducing flowering.
4. **Cytokinins**: Cytokinins are primarily produced in the root meristems and developing seeds. They are involved in cell division and promote lateral shoot growth. Cytokinins also play a role in delaying senescence, ensuring the longevity of plant tissues.
5. **Abscisic Acid (ABA):** ABA is primarily synthesized in response to stress conditions, such as drought, and it plays a crucial role in regulating water balance. ABA induces stomatal closure to reduce water loss and is involved in seed dormancy and germination.
6. **Ethylene**: Ethylene is a gaseous hormone produced in response to stress, such as mechanical damage or ripening fruits. Ethylene promotes fruit ripening, senescence, and the abscission (shedding) of leaves and fruits.
7. **Brassinosteroids**: Brassinosteroids are involved in promoting cell expansion and division, as well as enhancing stress tolerance. They play a role in regulating various physiological processes, including flowering and root development.
8. **Hormonal Interactions**: Plant hormones do not act in isolation but interact with each other to fine-tune plant responses. For example, auxins and cytokinins work together to regulate cell division and differentiation during organ development. The balance between auxins and gibberellins influences stem elongation and internode length.
9. **Environmental Factors and Hormonal Regulation**: External environmental factors, such as light, temperature, and stress conditions, influence hormonal regulation in plants. For instance, light exposure affects the synthesis and distribution of auxins, impacting phototropism and growth direction.

Hormonal regulation is a complex and dynamic process that underlies plant growth, development, and responses to environmental cues. Understanding the roles of different plant hormones and their interactions is essential for horticulturists and plant biologists to manipulate plant growth and optimize crop yield. The precise control of hormonal regulation offers opportunities for sustainable agriculture and the development of stress-tolerant crop varieties to meet the challenges of a changing climate and growing global population.

Soil Composition and Properties
Soil is a complex and dynamic natural resource that plays a crucial role in supporting plant growth and sustaining life on Earth. It is a product of long-term interactions between geological, biological, and environmental factors. Understanding soil composition and its properties is fundamental to effective soil management and optimizing crop production. In this section, we will delve into the details of soil composition and the various properties that influence its fertility and overall health.

Soil Components:

Soil is composed of four major components: minerals, organic matter, water, and air. Each component plays a vital role in soil fertility and productivity.

1. **Minerals**: Minerals are the inorganic particles derived from the weathering of rocks and minerals in the Earth's crust. The most common soil minerals are clay, silt, and sand. Clay particles are the smallest, followed by silt and sand. The proportion of these particles in the soil determines the soil texture, which greatly influences its water-holding capacity, aeration, and nutrient availability.
2. **Organic Matter**: Organic matter in the soil is composed of decaying plant and animal residues, along with microorganisms. It serves as a significant source of nutrients for plants, helps improve soil structure, enhances water retention, and promotes beneficial microbial activity. Organic matter also aids in reducing soil erosion and increases the soil's ability to withstand environmental stresses.
3. **Soil Horizons:**

Soil is typically arranged in distinct layers called soil horizons, each with unique properties and characteristics. The three main horizons are:

O Horizon: The topmost layer consists of organic matter, such as decomposing leaves and plant debris. It is commonly found in forested areas and acts as a source of nutrients for the underlying horizons.
A Horizon (Topsoil): This layer is rich in organic matter and is the primary zone of root activity. It contains a mix of mineral particles, organic material, and microorganisms, making it the most fertile layer of the soil.
B Horizon (Subsoil): The subsoil is characterized by the accumulation of minerals leached from the A horizon. It contains fewer organic materials but may have accumulated minerals like iron and clay. The subsoil often acts as a reservoir of nutrients that can be accessed by plants during periods of limited nutrient availability.
C Horizon (Parent Material): This layer contains weathered rocks and minerals and serves as the parent material for the development of upper soil horizons. It is less affected by biological processes and lacks the distinct properties of the upper horizons.

Soil Properties:

Several essential properties influence soil health and its ability to support plant growth. These properties include:

1. **Soil Texture**: Soil texture refers to the proportion of sand, silt, and clay particles in the soil. Sandy soils have larger particles, offering good drainage but limited water and nutrient retention. Clay soils, with their fine particles, have excellent water and nutrient retention but poor drainage. Loam, a balanced combination of sand, silt, and clay, is considered ideal for plant growth due to its optimal water-holding capacity and aeration.
2. **Soil Structure**: Soil structure refers to the arrangement of soil particles into aggregates or crumbs. A well-structured soil has a favorable crumb size, allowing for good root penetration, water movement, and air circulation. Proper soil structure promotes root development and provides a suitable environment for beneficial soil organisms.
3. **Soil pH:** Soil pH is a measure of the soil's acidity or alkalinity on a scale of 0 to 14, with 7 being neutral. Most plants thrive in slightly acidic to neutral soils (pH 6 to 7). Soil pH significantly influences nutrient availability; for example, some nutrients become less available to plants in alkaline soils.
4. **Soil Moisture**: Soil moisture is the amount of water present in the soil. Adequate soil moisture is essential for plant growth and nutrient uptake. However, excess water can lead to oxygen deficiency in the root zone, causing root rot and other issues. Proper irrigation management is crucial to maintaining optimal soil moisture levels.
5. **Soil Porosity:** Soil porosity refers to the amount of pore spaces or gaps between soil particles. These spaces allow for the movement of air and water through the soil, facilitating root respiration and nutrient uptake. Compacted soils have reduced porosity, which can restrict root growth and water infiltration.
6. **Cation Exchange Capacity (CEC):** CEC is a measure of the soil's ability to hold and exchange cations (positively charged ions) like calcium, potassium, and magnesium with the soil solution. Soils with higher CEC can retain more nutrients, reducing the risk of nutrient leaching and increasing nutrient availability to plants.

The composition and properties of soil is essential for effective soil management. Soil testing and analysis can provide valuable insights into the soil's fertility status, enabling farmers and horticulturists to make informed decisions about nutrient application and other soil management practices. By optimizing soil health, we can promote sustainable agriculture, enhance crop productivity, and preserve the health of our ecosystems.

Soil Formation and Classification

Soil formation is a complex and continuous natural process influenced by geological, climatic, biological, and topographic factors. Over time, the interaction of these factors leads to the development of distinct soil profiles and characteristics. Understanding the process of soil formation and classifying soils based on their properties is essential for effective soil management and sustainable agricultural practices. In this section, we will explore the intricate process of soil formation and the classification systems used to categorize different soil types.

Soil Formation:

Soil formation, also known as pedogenesis, begins with the weathering of rocks and minerals. Physical, chemical, and biological weathering processes break down parent materials into smaller particles, contributing to the formation of the C horizon—the parent material layer in soil profiles. Subsequently, environmental factors such as climate, topography, and time come into play, influencing the development of soil horizons.

1. **Climate:** Climate is a critical factor in soil formation, as it influences the rate and intensity of weathering. For instance, in regions with high rainfall, leaching can remove soluble nutrients from the upper soil horizons, leading to nutrient-poor soils. In contrast, arid regions may have soils with high mineral content due to limited leaching and weathering.
2. **Parent Material**: The composition of the parent material determines the initial properties of the soil, including texture, mineral composition, and nutrient content. Different rocks and minerals weather at varying rates, resulting in soils with distinct characteristics.
3. **Topography**: The landscape's shape and elevation influence factors like water drainage, erosion, and deposition, which, in turn, impact soil development. Slopes and hillsides may have thinner soils due to erosion, while valleys may accumulate fertile alluvial soils from sediment deposition.
4. **Biological Activity**: The activities of plants, animals, and microorganisms greatly influence soil development. Plant roots can physically break up rocks and contribute organic matter to the soil through litterfall and root decay. Soil organisms, such as earthworms and bacteria, play a crucial role in decomposing organic matter and releasing nutrients.
5. **Time**: Soil formation is a slow process that occurs over thousands of years. Older soils tend to have well-developed horizons, whereas younger soils may have less distinct layers.

Soil Classification:

Soil classification is the process of categorizing soils into groups based on their properties and characteristics. Various soil classification systems exist, with the most commonly used being the Soil Taxonomy developed by the United States Department of Agriculture (USDA). This system classifies soils based on diagnostic properties and hierarchical categories.

Order: The highest level of classification, the soil order, is based on dominant soil-forming processes and the presence of specific diagnostic horizons. Major soil orders include Entisols, Inceptisols, Mollisols, Alfisols, Ultisols, Aridisols, Spodosols, and Oxisols.

Suborder: Suborders are subdivisions of soil orders and reflect further differentiation based on regional variations or minor differences in soil properties.

Great Group: Great groups are distinguished by additional diagnostic properties and characteristics within a suborder.

Subgroup: Subgroups represent more specific variations of great groups, usually based on soil moisture, temperature, or parent material.

Family: Families are based on properties related to soil horizons, mineralogy, or drainage.

Series: The finest level of classification, the soil series, represents soils with similar characteristics and properties within a family.

Soil classification helps scientists, farmers, and land managers understand the capabilities and limitations of soils for specific land uses. It aids in selecting appropriate crops, determining irrigation and nutrient management practices, and identifying soil conservation measures. Accurate soil classification is vital for sustainable land use planning and responsible environmental stewardship.

Soil Fertility and Nutrient Management

Soil fertility refers to the soil's ability to provide essential nutrients in adequate quantities to support plant growth and maintain healthy ecosystems. Soil fertility is critical for sustaining agricultural productivity and ensuring food security. Effective nutrient management is a crucial aspect of soil fertility, as it involves supplying essential nutrients to plants in a balanced and sustainable manner. In this section, we will explore soil fertility, the role of essential nutrients, and various nutrient management practices to optimize plant growth and yield.

Essential Nutrients for Plant Growth:

Plants require a range of essential nutrients for proper growth and development. These nutrients can be broadly classified into two categories:

1. **Macronutrients:** Macronutrients are required in relatively large quantities by plants. They include nitrogen (N), phosphorus (P), potassium (K), calcium (Ca), magnesium (Mg), and sulfur (S). Nitrogen is essential for leafy growth and chlorophyll production, phosphorus promotes root development and flower formation, potassium is critical for overall plant vigor and disease resistance, and calcium and magnesium contribute to cell structure and enzyme function.
2. **Micronutrients:** Micronutrients are essential elements required in trace amounts. They include iron (Fe), manganese (Mn), zinc (Zn), copper (Cu), boron (B), molybdenum (Mo), and chlorine (Cl). Although plants need these nutrients in small quantities, they are equally vital for various physiological processes, such as enzyme activation and hormone synthesis.

Soil Nutrient Management:

Maintaining soil fertility and nutrient availability is essential for sustaining crop yields and minimizing nutrient losses. Several nutrient management practices help optimize nutrient uptake by plants and reduce the risk of nutrient leaching or runoff:

1. **Soil Testing:** Regular soil testing is essential for assessing the nutrient status of the soil. Soil samples are analyzed to determine the pH, nutrient levels, and other properties, allowing farmers to tailor nutrient management strategies based on crop requirements and existing soil conditions.
2. **Fertilizer Application:** Fertilizers are used to supplement the soil with essential nutrients. The choice of fertilizers and their application rates depend on the crop type, soil nutrient levels, and growth stage of the plants.
3. **Organic Matter Incorporation**: Adding organic matter to the soil improves soil structure,

enhances nutrient holding capacity, and promotes microbial activity. Organic matter sources include compost, manure, cover crops, and crop residues.

4. **Crop Rotation:** Crop rotation involves alternating different crop types on the same piece of land over successive growing seasons. This practice helps balance nutrient demands, reduces the build-up of pests and diseases, and improves soil health.

5. **Cover Crops:** Planting cover crops during fallow periods or between main crop rotations helps prevent soil erosion, adds organic matter to the soil, and fixes nitrogen from the atmosphere, making it available for subsequent crops.

6. **Nutrient Timing:** Proper timing of nutrient application is crucial for maximizing nutrient uptake by plants. Nutrients should be applied when plants have the highest demand, such as during the active growing season.

7. **Controlled Release Fertilizers:** These fertilizers release nutrients gradually over time, matching the nutrient supply with the crop's demand, reducing nutrient losses, and enhancing efficiency.

8. **Irrigation Management:** Proper irrigation practices can minimize nutrient leaching by preventing excessive water application, which can carry nutrients below the root zone.

Optimizing soil fertility and nutrient management ensures sustainable agricultural practices, reduces environmental impacts, and contributes to long-term soil health. By adopting these practices,

Soil Erosion and Conservation

Soil erosion is a natural process that has been accelerated by human activities, resulting in the loss of valuable topsoil and threatening agricultural productivity and environmental stability. Soil conservation is the practice of preventing or reducing soil erosion and preserving the integrity of this precious natural resource. In this section, we will delve into the details of soil erosion, its causes, and various soil conservation techniques to protect soil health and promote sustainable land management.

Soil Erosion:

Soil erosion is the process by which soil particles are detached, transported, and deposited by the action of wind, water, or human activities. Natural erosion occurs slowly over time, but human interventions, such as deforestation, improper land use, and excessive agricultural practices, can accelerate erosion rates significantly. The primary agents of soil erosion are:

1. **Water Erosion**: Water erosion occurs when rainfall or irrigation water flows over the soil surface, carrying away soil particles. It can take the form of sheet erosion (thin layers of soil removed uniformly), rill erosion (small channels forming on slopes), or gully erosion (deep and wide channels resulting from concentrated flow).

2. **Wind Erosion:** Wind erosion is prevalent in arid and semi-arid regions where dry and loose soils are susceptible to wind action. Wind carries away fine soil particles, creating dust storms and reducing soil fertility.

Causes of Soil Erosion:

Human activities significantly contribute to soil erosion rates. Some of the key causes include:

1. **Deforestation:** The removal of natural vegetation exposes the soil to the impact of rainfall and wind, leading to increased erosion rates.
2. **Overgrazing:** Excessive grazing by livestock can strip vegetation cover, leaving the soil vulnerable to erosion.
3. **Unsustainable Agricultural Practices**: Poor land management, such as excessive tillage, improper irrigation, and lack of cover crops, can exacerbate erosion on agricultural lands.
4. **Construction and Urbanization:** Construction activities and urban development remove vegetation and disturb soil, leading to increased erosion.

Soil Conservation Techniques:

Soil conservation aims to prevent or reduce soil erosion and protect the soil's fertility and structure. Various conservation techniques are employed to safeguard soil health:

1. **Terracing:** On steep slopes, constructing terraces helps reduce the flow of water, slowing its velocity and minimizing erosion. Terracing creates flat areas that allow for more efficient water infiltration and reduce soil runoff.
2. **Contour Farming:** Planting crops along the contours of the land rather than straight up and down slopes helps minimize water flow and soil movement. Contour farming slows the water's speed, allowing it to infiltrate into the soil and minimizing erosion.
3. **Cover Crops:** Planting cover crops, such as legumes and grasses, during fallow periods or between main crop rotations helps protect the soil surface from the impact of raindrops, wind, and surface runoff.
4. **Mulching:** Applying organic or inorganic mulch materials on the soil surface helps reduce water runoff, control soil temperature, and prevent soil particle detachment by raindrops.
5. **Conservation Tillage:** Reducing or eliminating tillage helps maintain crop residue on the soil surface, reducing erosion and promoting soil organic matter accumulation.
6. **Windbreaks:** Planting rows of trees or shrubs perpendicular to the prevailing wind direction creates windbreaks that reduce wind speed and protect soils from wind erosion.
7. **Contour Buffer Strips:** Planting grass or other vegetation along the contour of sloping land helps slow water flow and trap sediment, preventing gully formation.
8. **Streambank Stabilization:** Erosion control measures along riverbanks and water bodies, such as riparian vegetation planting and structural measures, help prevent sediment loss into watercourses.
9. **Sediment Basins:** Constructing sediment basins traps sediment-laden runoff, allowing water to settle and preventing sediment from entering downstream water bodies.
10. **Education and Awareness:** Promoting awareness and education about soil conservation practices helps foster a culture of responsible land stewardship and encourages adoption of sustainable soil management practices.

By implementing these soil conservation techniques, we can reduce soil erosion rates, conserve soil fertility, enhance water quality, and preserve the health of ecosystems. Soil conservation is vital for achieving sustainable agriculture and ensuring the continued availability of fertile soils for future generations.

Chapter 7:
Nursery Management for Horticultural Crops

Nursery management is a crucial aspect of horticulture that involves the production and care of young plants, commonly known as seedlings or saplings, before they are transplanted to their final growing locations. This chapter provides a comprehensive overview of nursery management practices and explores the various types of nurseries used for horticultural crops.

Principles of Nursery Management:

➢ Seed Selection and Germination:

Importance of selecting high-quality seeds for nursery production. Proper seed storage and germination techniques are emphasized to ensure uniform and vigorous seedling establishment.

➢ Growing Media and Containers:

The choice of growing media and containers is critical in nursery management. The chapter explores different growing media options, such as peat-based mixes, compost, and coco coir, and discusses the advantages of using various container types, including pots, trays, and plug trays.

➢ Irrigation and Water Management:

Proper watering practices are vital for successful nursery management. The chapter delves into various irrigation methods, including drip irrigation and overhead irrigation, and highlights the importance of efficient water management to prevent waterlogging or dehydration.

➢ Temperature and Environmental Control:

Maintaining optimal temperature and humidity levels in nurseries is essential for seedling growth and health. The chapter explores techniques such as shade houses, misting systems, and ventilation to create favorable nursery conditions.

➢ Pest and Disease Management:

challenges of pests and diseases in nurseries and discusses integrated pest management (IPM) strategies to minimize the impact on young plants. Topics include biological controls, cultural practices, and pesticide application guidelines.

➢ Fertilization:

The role of balanced fertilization in nursery management is highlighted. The chapter discusses the use of organic and inorganic fertilizers and emphasizes the importance of providing essential nutrients for optimal seedling growth.

> Transplanting:

o transplanting seedlings from nurseries to their final growing locations. Proper techniques to minimize transplant shock and ensure a smooth transition are discussed.

Different Types of Nurseries for Horticultural Crops:

> Seedling Nursery:

seedling nurseries, which are dedicated to germinating seeds and nurturing young seedlings until they are ready for transplantation. The diverse range of horticultural crops suited for seedling nurseries is explored.

> Grafting Nursery:

grafting nurseries, where specialized techniques are used to join rootstock and scion materials to create grafted plants with desired traits.

> Tissue Culture Nursery:

Tissue culture nurseries employ sterile laboratory conditions to propagate horticultural plants from small tissue samples, enabling the production of genetically identical clones on a large scale.

> Cutting Nursery:

cutting nurseries, which propagate plants from stem cuttings, allowing for the cloning of parent plants with specific characteristics.

> Container Nursery:

Container nurseries use containers like pots or trays to grow young plants, providing better control over the growing environment and facilitating efficient transplanting.

> Bare-root Nursery:

bare-root nurseries, where young plants are grown and stored without soil around their roots, making them easier to transport and transplant.

> Shade House Nursery:

Shade house nurseries create a controlled microclimate, protecting young plants from excessive sunlight and temperature fluctuations.

- ➢ Greenhouse Nursery:

Greenhouse nursery which provide a controlled environment for seedlings, extending the growing season and protecting plants from adverse weather.

- ➢ Hardening Nursery:

  Hardening nursery which prepare seedlings for outdoor conditions before transplanting, reducing transplant shock and improving survival rates.

Nursery management and the various types of nurseries are vital components of successful horticultural crop production. This chapter equips readers with the knowledge and practices needed to establish and maintain healthy and robust nurseries, ensuring the supply of high-quality seedlings for sustainable horticulture.

Chapter 8

Growing Green: Methods of Plant Propagation

Plant propagation is the process of creating new plants from existing ones, enabling the multiplication and distribution of desirable plant varieties. There are two primary methods of plant propagation: sexual and asexual. Sexual propagation involves the use of seeds, which carry genetic information from both parent plants and result in genetic diversity. Asexual propagation, on the other hand, involves reproducing plants without seeds, typically through vegetative means such as cuttings, grafting, layering, or tissue culture.

Seed Propagation: Seed propagation is one of the most common and efficient methods of propagating plants.

Seed germination is the process in which a seed transforms into a young seedling, capable of producing a new plant. Germination begins when a seed absorbs water, leading to the activation of enzymes and the resumption of metabolic activities. The following stages are involved in seed germination:

1. **Water Absorption**: The first step in germination is the imbibition of water by the seed. As the seed absorbs water, it swells, softens the seed coat, and activates enzymes.
2. **Respiration:** With the availability of water and oxygen, respiration begins, converting stored nutrients within the seed into energy for growth.
3. **Cell Division and Elongation:** The embryo inside the seed begins to grow, and the cells divide and elongate, forming a small root (radicle) and a shoot (plumule).
4. **Emergence:** Once the radicle breaks through the seed coat, the seedling emerges from the soil and continues to grow into a mature plant.

Seed Viability:

Seed viability refers to the ability of a seed to germinate and produce a healthy seedling under favorable conditions. Not all seeds are viable, as viability decreases over time due to factors such as age, improper storage, and exposure to unfavorable conditions. Testing seed viability is

crucial before sowing, as it ensures efficient use of resources and prevents unnecessary efforts on non-viable seeds. Several methods can be used to assess seed viability:

1. **Germination Test:** The most common method, the germination test, involves sowing a representative sample of seeds under optimal conditions and counting the number of seeds that successfully germinate.
2. **Tetrazolium Test:** This staining test uses tetrazolium chloride to differentiate viable embryos (red or pink color) from non-viable ones (remains colorless).
3. **Electrical Conductivity Test:** This test measures the leachate's electrical conductivity obtained by soaking seeds, which correlates with seed viability.
4. **X-ray Test:** X-ray imaging allows the visualization of the embryo's internal structures, providing insight into its viability and development stage.

Seed Storage:

Proper seed storage is vital to maintain seed viability and prolong their shelf life. Seeds are living organisms with limited energy reserves, and they can lose viability if exposed to unfavorable conditions. To preserve seed quality during storage, the following factors should be considered:

1. **Moisture Content:** Seeds should be dried to a suitable moisture content to prevent germination and the growth of molds or fungi. A low moisture content (usually around 5-8%) is ideal for long-term storage.
2. **Temperature:** Seeds should be stored at cool temperatures to reduce metabolic activity and slow down aging processes. The recommended storage temperature varies for different plant species but generally ranges from 0°C to 10°C.
3. **Light:** Seeds should be stored in the dark or opaque containers to avoid light-induced germination and to prevent photodegradation of seed components.
4. **Airtight Containers:** Storing seeds in airtight containers protects them from exposure to oxygen and prevents moisture fluctuations.
5. **Desiccants:** The use of desiccants, such as silica gel, can help absorb excess moisture in the storage container, further protecting seeds from fungal growth.
6. **Periodic Testing:** It is advisable to periodically test seed viability during storage to assess the remaining germination potential and to renew or replace older seeds as needed.

The process of seed germination, assessing seed viability, and following proper seed storage practices, horticulturists and farmers can ensure a successful seed propagation process. Seed propagation plays a crucial role in plant breeding, crop production, and ecosystem restoration, contributing to the sustainable cultivation of a wide variety of plant species.

Vegetative Propagation: Cuttings, Grafting, Layering, and Budding

Vegetative propagation is a method of plant propagation that involves the use of plant parts other than seeds to create new individuals with identical genetic characteristics to the parent plant. This technique is widely used in horticulture to propagate desirable plant varieties and maintain their unique traits. In this section, we will explore the four primary methods of vegetative propagation: cuttings, grafting, layering, and budding, including their procedures, advantages, and applications.

Cuttings:

Cuttings involve taking a portion of a stem, leaf, or root from a parent plant and encouraging it to develop roots and shoots to form a new plant. There are three main types of cuttings:

1. **Stem Cuttings:** Stem cuttings involve taking a piece of stem with a few nodes (where leaves attach) and removing the lower leaves. The cutting is then inserted into a growing medium, and under favorable conditions, it will develop roots and shoots.
2. **Leaf Cuttings:** Leaf cuttings are taken from leaves with a portion of the petiole (leaf stalk) and

are usually placed horizontally on the growing medium. New plantlets will develop from the base of the leaf.

3. **Root Cuttings:** Root cuttings involve taking sections of roots and planting them in a growing medium, where they will produce new shoots.

Cuttings are suitable for many plant species, including shrubs, herbaceous plants, and some woody plants. They allow for the rapid multiplication of desired varieties and the preservation of specific traits.

Grafting:

Grafting is a technique in which a stem (scion) from one plant is joined to the root system (rootstock) of another plant. This method allows for the combination of desirable qualities from two different plants. Grafting is commonly used for fruit trees, ornamental plants, and some woody plants. The main types of grafting include:

1. **Whip and Tongue Grafting:** In this method, both the scion and rootstock are cut diagonally, and a small tongue-like incision is made in each. The scion and rootstock are then joined together, allowing for a close union.
2. **Cleft Grafting:** A vertical split is made in the rootstock, and the scion, with a corresponding wedge shape, is inserted into the cleft.
3. **Bark Grafting:** The scion is inserted under the bark of the rootstock, where it will fuse and grow.
4. **Budding:** Budding is a form of grafting in which a single bud, rather than a whole stem, is inserted into the rootstock. The bud will develop into a new shoot.

Grafting allows for the propagation of rare or difficult-to-root plants, the rejuvenation of old or diseased trees, and the combination of desirable characteristics in a single plant.

Layering:

Layering is a method of vegetative propagation in which a branch or stem is encouraged to produce roots while still attached to the parent plant. There are different types of layering:

1. **Simple Layering:** A low-growing branch is bent to the ground and partially buried, allowing it to develop roots. Once rooted, it can be separated from the parent plant and transplanted.
2. **Air Layering:** In air layering, a portion of the stem is girdled to disrupt the flow of nutrients. The area is then covered with moist soil or a rooting medium, and roots will form above the girdle.

Layering is a useful method for propagating plants that may be difficult to root from cuttings or for producing new plants that are already established and larger in size.

Budding:

Budding is a technique in which a single bud from a plant (the scion) is inserted into a slit or T-shaped cut on the bark of another plant (the rootstock). Budding is commonly used for fruit trees, roses, and ornamental plants. The main types of budding include:

1. **T-Budding:** In T-budding, a T-shaped cut is made on the rootstock, and a single bud from the scion is inserted into the T-cut.
2. **Chip Budding**: In chip budding, a small chip of the scion, including a bud, is inserted into a matching cut on the rootstock.

Budding allows for the efficient propagation of desired varieties, especially in fruit tree production, where specific fruit characteristics can be maintained.

Vegetative propagation methods, such as cuttings, grafting, layering, and budding, offer valuable tools for horticulturists to propagate and preserve desired plant varieties with consistent genetic traits. These techniques are essential in the production of new plants with desired characteristics and the preservation of unique and valuable plant germplasm. Each method has its advantages and applications, and by mastering these techniques, horticulturists can create and maintain diverse and thriving plant populations.

Micropropagation and Tissue Culture Techniques

Micropropagation and tissue culture techniques are advanced methods of plant propagation that involve the cultivation of plant cells, tissues, or organs in a controlled laboratory environment. These techniques allow for the rapid multiplication of elite plant varieties, the production of

disease-free plants, and the preservation of endangered or rare species. In this section, we will explore the principles and procedures of micropropagation and tissue culture, along with their applications in horticulture and plant biotechnology.

Principles of Micropropagation and Tissue Culture:

Micropropagation and tissue culture are based on the principles of plant cell totipotency, the ability of plant cells to dedifferentiate and develop into complete plants under appropriate conditions. The key steps in micropropagation and tissue culture include:

1. Initiation: Plant material, such as a piece of stem or leaf, is collected from the parent plant and sterilized to remove any microorganisms.
2. Establishment: The sterilized plant material is placed onto a nutrient-rich culture medium containing essential nutrients, growth hormones, and carbohydrates, providing an environment for the cells to divide and grow.
3. Multiplication: Under controlled conditions, the cells divide and form callus, a mass of undifferentiated cells. The callus is then subcultured onto fresh media to promote rapid multiplication.
4. Shoot Formation: From the callus, shoots are induced by manipulating the concentration of growth hormones in the culture medium.
5. Rooting: Once the shoots have developed, they are transferred to a rooting medium to induce the formation of roots.

6. Acclimatization: The rooted plantlets are gradually acclimatized to external environmental conditions in a controlled greenhouse or nursery before being transferred to the field.

Applications of Micropropagation and Tissue Culture:

Micropropagation and tissue culture techniques have numerous applications in horticulture, plant biotechnology, and agriculture:

1. Mass Propagation: Micropropagation allows for the rapid production of a large number of genetically identical plants within a short period, providing a valuable tool for commercial plant production.
2. Disease Elimination: Tissue culture techniques can be used to produce disease-free plants from infected plant material, contributing to plant health and improved crop yields.
3. Clonal Selection: Elite varieties with desirable traits, such as high yield, disease resistance, or specific growth habits, can be efficiently multiplied and preserved through tissue culture.
4. Genetic Engineering: Tissue culture serves as a foundation for genetic engineering and biotechnology applications, where specific genes can be introduced or modified in plant cells.
5. Conservation of Endangered Species: Rare or endangered plant species can be preserved and propagated through tissue culture to prevent extinction and preserve genetic diversity.
6. Hybridization: In breeding programs, tissue culture techniques facilitate the production of interspecific and intergeneric hybrids, overcoming barriers in traditional cross-breeding.
7. Crop Improvement: Through tissue culture, scientists can manipulate the genetic makeup of plants to enhance desired traits, such as improved nutritional content or drought tolerance.

Challenges of Micropropagation and Tissue Culture:

Despite the numerous benefits, micropropagation and tissue culture techniques also present certain challenges:

1. Contamination: Maintaining a sterile environment is critical to prevent contamination by bacteria, fungi, or other microorganisms that can hinder tissue culture success.
2. Genotypic Variation: Slight differences in growth conditions or hormonal treatments can result in variations among the propagated plants, requiring rigorous selection and testing.
3. Cost: The initial set-up and maintenance costs of tissue culture laboratories can be substantial, especially for small-scale operations.
4. Skill Requirements: Tissue culture techniques require specialized skills and training to achieve consistent and successful results.

Micropropagation and tissue culture techniques have revolutionized plant propagation and are indispensable tools in modern horticulture and plant biotechnology. These methods enable the efficient production of a wide range of plant species with desirable traits, contributing to sustainable agriculture, conservation efforts, and the advancement of plant science.

Chapter 9:

Crops Production & Management

Principles of Crop Production

Horticultural crop production involves the cultivation of fruits, vegetables, flowers, and ornamental plants for human consumption, aesthetics, and various industrial uses. To achieve successful crop production, horticulturists must understand and implement fundamental principles that optimize plant growth, yield, and quality.

Soil Preparation:

The foundation of successful crop production lies in the preparation of the soil, ensuring that it provides the necessary nutrients, aeration, and water retention for plant growth. The key principles of soil preparation include:

1. Soil Testing: Before planting, conducting a soil test is essential to assess the soil's nutrient content and pH level. The results help determine the appropriate fertilizers and amendments required to optimize plant growth.
2. Soil Fertility: Adding organic matter, such as compost or well-rotted manure, improves soil structure and fertility, enhancing nutrient availability to plants.
3. Soil Aeration: Proper soil aeration is crucial for root development and oxygen supply to plant roots. Avoiding soil compaction and incorporating organic matter can improve soil aeration.
4. Drainage: Ensuring adequate drainage is vital to prevent waterlogged conditions, which can lead to root rot and other plant diseases. Proper grading and installation of drainage systems can help manage excess water.
5. Soil pH Adjustment: Some crops have specific pH requirements for optimal growth. Adjusting the soil pH using lime or sulfur can create a suitable environment for the selected crops.Crop Selection:

Choosing the right crop for the given climate, soil type, and available resources is essential for successful crop production. Considerations for crop selection include:

1. Climate Suitability: Different crops have varying temperature and moisture requirements. Selecting crops suitable for the local climate helps ensure higher yields and reduced risk of crop failure.
2. Soil Type: Some crops thrive in specific soil types, such as sandy, loamy, or clay soils. Selecting crops that are well-adapted to the soil type prevents growth challenges.
3. Market Demand: Considering market demand for specific crops is essential to ensure profitability and marketing opportunities.
4. Crop Rotation: Implementing a crop rotation system helps maintain soil fertility, reduce pest and disease buildup, and minimize the risk of crop-specific nutrient deficiencies.

Planting Techniques:

Proper planting techniques are critical to promoting strong seedling establishment and early growth. Key principles of planting techniques include:

1. Plant Spacing: Proper spacing ensures adequate sunlight, airflow, and nutrient availability for each plant, reducing competition and maximizing individual plant growth.
2. Planting Depth: Planting seeds or seedlings at the correct depth ensures that the emerging roots have access to essential nutrients and moisture.
3. Timing: Planting at the appropriate time, considering the local climate and growing season, optimizes crop development and yield.
4. Irrigation: Providing sufficient water during the initial establishment phase helps young plants establish healthy root systems.

Crop Management Practices:

Crop management practices encompass a range of activities designed to promote healthy plant growth, maximize yield, and minimize pest and disease pressure. Important crop management principles include:

1. Irrigation Management: Implementing efficient irrigation practices, such as drip irrigation or controlled irrigation scheduling, ensures proper water distribution and minimizes water wastage.
2. Nutrient Management: Providing balanced and appropriate nutrients to crops through fertilization promotes healthy growth and high yields.
3. Weed Control: Proper weed management is crucial to minimize competition for resources and reduce the spread of pests and diseases.
4. Pest and Disease Management: Monitoring for pests and diseases and implementing integrated pest management (IPM) strategies help reduce the use of chemical pesticides and minimize crop damage.
5. Pruning and Training: Pruning and training plants help improve airflow, light penetration, and fruit production, enhancing crop quality and yield.
6. Harvesting and Post-Harvest Handling: Harvesting at the right stage of maturity and proper post-harvest handling practices maintain crop quality and extend shelf life.

Applying the principles of crop production is essential for horticulturists to achieve successful, sustainable, and profitable crop cultivation. By understanding soil preparation, selecting appropriate crops, employing effective planting techniques, and implementing sound crop management practices, horticulturists can optimize plant growth and yield, resulting in high-quality produce and flourishing horticultural enterprises.

Planning and Establishment of Horticultural Crops

Planning and establishment are critical stages in horticultural crop production that lay the foundation for a successful and productive crop. Proper planning involves making informed decisions about crop selection, site preparation, and resource allocation. The establishment phase includes activities such as land preparation, planting, and early crop care. In this section, we will explore the key considerations and steps involved in planning and establishing horticultural crops.

Crop Selection and Site Assessment:

The first step in planning horticultural crop production is selecting suitable crops based on market demand, climatic conditions, and site suitability. Considerations for crop selection and site assessment include:

1. Market Demand: Identify crops with a high demand in the local or target market. Consider factors such as consumer preferences, market trends, and potential profitability.
2. Climate and Growing Conditions: Choose crops that are well-suited to the local climate, including temperature, rainfall, humidity, and sunlight requirements.
3. Soil Suitability: Assess the soil type, fertility, pH, and drainage to determine whether it is suitable for the selected crops. Consider soil testing and amendments if necessary.
4. Water Availability: Evaluate the water source and availability for irrigation, especially in regions with limited rainfall.
5. Pest and Disease Pressure: Consider potential pest and disease issues in the area and select crop varieties with natural resistance or develop pest management strategies.

Land Preparation:

Proper land preparation is essential to create an optimal environment for crop growth and establishment. Key steps in land preparation include:

1. Clearing and Grubbing: Remove existing vegetation, debris, and rocks from the site to create a clean planting area.
2. Soil Cultivation: Plow or till the soil to break it up and create a fine seedbed. This improves soil aeration, facilitates root penetration, and enhances water infiltration.
3. Leveling: Level the land to ensure uniform water distribution and prevent waterlogging or erosion.
4. Incorporating Organic Matter: Add organic matter, such as compost or well-rotted manure, to improve soil fertility and structure.

Planting:

Planting is a critical phase in horticultural crop production that requires attention to detail to ensure successful establishment. Important aspects of planting include:

1. Planting Method: Select the appropriate planting method based on the crop type and site conditions. Common planting methods include direct seeding, transplanting, and containerized planting.
2. Plant Spacing: Determine the optimal spacing between plants to allow for proper growth, light penetration, and airflow.
3. Planting Depth: Plant seeds or seedlings at the appropriate depth to ensure optimal root development.

4. Watering: Provide sufficient water at planting to aid seedling establishment and root development.

Early Crop Care:

In the early stages of crop growth, plants require careful attention to promote healthy development. Early crop care activities include:

1. Irrigation: Maintain adequate soil moisture through regular irrigation or efficient water management practices.
2. Weed Control: Control weeds to minimize competition for resources and prevent weed interference.
3. Fertilization: Apply appropriate fertilizers to provide essential nutrients for plant growth.
4. Pest and Disease Monitoring: Regularly inspect plants for signs of pests and diseases and implement early management measures if needed.
5. Mulching: Apply organic mulch around plants to conserve soil moisture, suppress weeds, and regulate soil temperature.

Proper planning and establishment of horticultural crops set the stage for successful crop production. By carefully selecting suitable crops, assessing site conditions, preparing the land, and implementing proper planting and early crop care practices, horticulturists can create a solid foundation for healthy and productive crops. Planning and establishment are integral components of sustainable horticultural practices, contributing to increased yields, enhanced crop quality, and overall success in the horticultural industry.

Pest, Disease, and Weed Management of Horticultural Crops

Pests, diseases, and weeds are significant challenges in horticultural crop production that can lead to reduced yields and quality. Integrated Pest Management (IPM) is an effective and sustainable approach to managing these issues while minimizing environmental impacts. IPM combines various pest management strategies and techniques to achieve long-term pest control and promote overall crop health. In this section, we will explore the principles and components of IPM and how they can be applied to effectively manage pests, diseases, and weeds in horticultural crops.

Principles of Integrated Pest Management (IPM):

IPM is based on several fundamental principles:

1. Pest Monitoring: Regularly monitor horticultural crops to detect the presence and abundance of pests, diseases, and weeds. Early detection allows for timely intervention.
2. Prevention: Implement preventative measures to reduce pest, disease, and weed pressure. This includes using disease-resistant crop varieties, employing good cultural practices, and establishing physical barriers.
3. Biological Control: Utilize natural enemies, such as beneficial insects, parasites, and predators,

to control pest populations. Encouraging biodiversity in the agroecosystem enhances natural pest control.

4. Cultural Practices: Adopt cultural practices, such as crop rotation, intercropping, and mulching, to disrupt pest life cycles and create unfavorable conditions for pests and diseases.
5. Chemical Control as a Last Resort: Chemical pesticides should be used only when non-chemical methods are not effective or practical. Selective and targeted pesticide use minimizes harm to beneficial organisms and the environment.

Components of Integrated Pest Management (IPM):

IPM involves the integration of multiple pest management components, which work synergistically to achieve effective pest control. The key components of IPM include:

1. Biological Control: Encouraging natural predators and parasites to reduce pest populations. This can involve releasing beneficial insects, conserving natural enemies, or introducing biological control agents.
2. Cultural Control: Implementing cultural practices that promote healthy plant growth and reduce pest, disease, and weed pressure. This includes crop rotation, proper irrigation, and sanitation.
3. Mechanical and Physical Control: Using physical methods to physically remove pests, such as handpicking insects or using barriers to prevent pest access to crops.
4. Chemical Control: When necessary, applying pesticides selectively and judiciously, using low-toxicity and targeted products to minimize impacts on beneficial organisms and the environment.
5. Monitoring and Thresholds: Regularly monitoring pest populations and applying control measures only when pest populations exceed predetermined economic or ecological thresholds.

Implementation of Integrated Pest Management (IPM) in Horticultural Crops:

The successful implementation of IPM in horticultural crops involves a systematic and holistic approach:

1. Identify Pest, Disease, and Weed Problems: Accurate identification of pests, diseases, and weeds is essential to design appropriate control strategies.
2. Monitor Pest Populations: Regularly scout fields and monitor pest populations to assess the need for intervention.
3. Establish Thresholds: Determine economic or ecological thresholds beyond which action is required to prevent yield or quality losses.
4. Implement Preventative Measures: Adopt preventative practices, such as crop rotation, sanitation, and the use of disease-resistant varieties, to reduce pest pressure.
5. Introduce Biological Control: Introduce natural enemies or biological control agents to suppress pest populations.
6. Use Chemical Control as a Last Resort: If necessary, use chemical pesticides judiciously and according to label instructions.
7. Evaluate and Adapt: Continuously assess the effectiveness of IPM strategies and make adjustments as needed to improve pest management.

Integrated Pest Management is an essential approach in horticultural crop production, promoting sustainable and environmentally friendly pest, disease, and weed management. By incorporating multiple strategies and practices, horticulturists can effectively manage pests and diseases while maintaining crop health and productivity in a balanced and eco-friendly manner.

Identification and Control of Common Pests and Diseases

Effective pest and disease management is crucial for maintaining healthy and productive horticultural crops. Timely identification and appropriate control measures are essential to prevent yield losses and ensure crop quality. In this section, we will explore common pests and diseases that affect horticultural crops and the strategies for their identification and control.

Identification of Common Pests and Diseases:

1. Insects: Insects are common pests that can cause damage to horticultural crops through feeding or transmitting diseases. Common insect pests include aphids, caterpillars, beetles, thrips, and mites.
2. Diseases: Horticultural crops are susceptible to various diseases caused by fungi, bacteria, viruses, and nematodes. Common diseases include powdery mildew, downy mildew, bacterial spot, leaf blights, and root rots.
3. Nematodes: Nematodes are microscopic worm-like organisms that feed on plant roots, causing stunted growth and reduced yields.
4. Weeds: Weeds are unwanted plants that compete with horticultural crops for nutrients, water, and sunlight. Common weed species include annual and perennial grasses, broadleaf weeds, and sedges.

Pest and Disease Control Strategies:

Effective pest and disease control strategies are essential for minimizing crop losses and ensuring healthy plant growth. The key control measures include:

1. Biological Control: Introducing natural enemies, such as predatory insects or parasites, to control pest populations. This biological control method is environmentally friendly and helps maintain a natural balance in the ecosystem.
2. Cultural Control: Implementing cultural practices to reduce pest and disease pressure. These practices include crop rotation, sanitation, proper irrigation, and planting disease-resistant crop varieties.
3. Chemical Control: Using pesticides as a last resort when other control measures are not sufficient. Selective and targeted pesticide use helps minimize the impact on beneficial organisms and the environment.
4. Integrated Pest Management (IPM): Integrating multiple pest and disease control strategies to achieve effective, sustainable, and environmentally friendly management. IPM combines biological, cultural, and chemical control methods to minimize pest and disease damage.

Weed Management Techniques:
Effective weed management is crucial to prevent weed competition and reduce the impact of weeds on horticultural crops. The key weed management techniques include:

1. Hand Weeding: Physically removing weeds by hand or with handheld tools. Hand weeding is suitable for small-scale or sensitive areas where chemical herbicides may not be appropriate.
2. Mulching: Applying organic or inorganic mulch around plants to suppress weed growth and conserve soil moisture. Mulching also helps regulate soil temperature and prevents weed seed germination.
3. Mechanical Control: Using mechanical tools, such as plows, cultivators, or mowers, to physically remove weeds. Mechanical control is suitable for larger areas but may require multiple passes to manage weeds effectively.
4. Chemical Control: Using herbicides to control weeds. Selective herbicides target specific weed species while leaving the horticultural crops unharmed. Non-selective herbicides kill all vegetation and are useful for clearing fields before planting.
5. Cover Crops: Planting cover crops that suppress weed growth and compete for resources with unwanted weed species.
6. Crop Rotation: Rotating crops to disrupt weed life cycles and prevent the buildup of specific weed species.
7. Biological Control: Introducing or encouraging weed-eating insects, such as goats or insects, to graze on weeds and keep their populations in check.

Effective identification and control of common pests, diseases, and weeds are essential components of successful horticultural crop management. By implementing appropriate control strategies and employing integrated pest and weed management techniques, horticulturists can ensure healthy and thriving crops, leading to improved yields and quality in the horticultural industry.

Harvesting, Post-Harvest Handling, and Storage of Horticultural Crops

Harvesting marks the culmination of efforts in horticultural crop production and represents the transition from growth to consumption or processing. After harvesting, proper post-harvest handling and storage practices are essential to maintain the quality and extend the shelf life of the crops. In this section, we will explore the key considerations and techniques involved in harvesting, post-harvest handling, and storage of horticultural crops.

Harvesting:

Harvesting is a critical activity that requires careful timing and attention to ensure the crops are picked at the peak of their quality and maturity. Key considerations for harvesting include:

1. Harvest Timing: Determine the optimal time for harvesting each crop based on factors such as size, color, taste, texture, and sugar content.
2. Selective Harvesting: When crops have different maturity stages, use selective harvesting to pick only ripe fruits or vegetables while leaving immature ones to continue ripening.
3. Harvesting Tools: Use appropriate harvesting tools, such as shears, knives, or clippers, to minimize damage to the crops during harvesting.
4. Handling Care: Handle harvested crops gently to avoid bruising and physical damage that can

lead to spoilage.

Post-Harvest Handling:

Post-harvest handling is a series of activities that aim to preserve the quality and freshness of the harvested crops from the field to the market or processing facility. Essential post-harvest handling practices include:

1. Cleaning: Remove dirt, debris, and any damaged or diseased parts from the harvested crops.
2. Cooling: Rapidly cool the crops to the appropriate storage temperature to slow down ripening and reduce the risk of decay.
3. Grading and Sorting: Sort the crops according to size, color, and quality to meet market demands and consumer preferences.
4. Packaging: Use suitable packaging materials to protect the crops during transportation and storage, while also providing proper ventilation to prevent moisture buildup.
5. Transportation: Handle and transport the harvested crops carefully to avoid damage and temperature fluctuations.

Storage:

Proper storage is crucial to prolong the shelf life and maintain the quality of horticultural crops. Considerations for storage include:

1. Storage Facilities: Use well-ventilated, clean, and pest-free storage facilities, such as cold rooms, warehouses, or refrigerated containers.
2. Temperature and Humidity: Set appropriate temperature and humidity levels for each crop to prevent spoilage and maintain freshness.
3. Ethylene Control: Ethylene, a natural ripening hormone, can accelerate the ripening of some fruits. Control ethylene levels in storage to extend shelf life.
4. Monitoring: Regularly monitor stored crops for signs of decay, disease, or pests, and remove any affected produce promptly.
5. Controlled Atmosphere Storage: Some crops benefit from controlled atmosphere storage, where oxygen, carbon dioxide, and ethylene levels are carefully managed to extend shelf life.

Shelf Life Extension Techniques:

To further extend the shelf life of horticultural crops, several techniques can be employed:

1. Pre-cooling: Pre-cool the crops immediately after harvesting to remove field heat and slow down the ripening process.
2. Post-Harvest Treatments: Use post-harvest treatments like waxing, hot water treatments, or modified atmosphere packaging to reduce moisture loss and slow down ripening.
3. Refrigeration and Controlled Atmosphere: Store crops in refrigeration or controlled atmosphere conditions to slow down physiological changes and reduce spoilage.

Proper harvesting, post-harvest handling, and storage practices are vital for minimizing losses, preserving the quality, and maximizing the market value of horticultural crops. By following these techniques, horticulturists can ensure that consumers receive fresh, nutritious, and visually appealing produce, contributing to the success and sustainability of the horticultural industry.

Chapter 10

Floriculture: A Horticultural Perspective

Floriculture is a specialized branch of horticulture dedicated to the cultivation and management of ornamental flowering plants for aesthetic and decorative purposes.

Importance of Floriculture in Horticulture

Floriculture holds immense importance in horticulture due to its significant contributions to the global ornamental plant industry. It plays a crucial role in beautifying landscapes, gardens, and public spaces, adding color, fragrance, and aesthetic appeal to the environment. Beyond the ornamental value, floriculture also contributes to the cultural, social, and economic aspects of society. Flowers have deep-rooted cultural significance and are used in various rituals, ceremonies, and celebrations worldwide. The cut flower and floral arrangement industry also serves as a lucrative market, generating substantial revenue and employment opportunities for horticulture practitioners.

Historical Evolution of Floriculture

The historical evolution of floriculture dates back to ancient civilizations, where ornamental gardens were developed for pleasure and spiritual enrichment. From the Persian and Islamic gardens to the European Renaissance gardens, each era contributed to the development of different styles and designs. The Victorian era saw the rise of floristry and the symbolic language of flowers, which continue to influence floral arrangements and cultural expressions today. As horticulture practices evolved, advancements in breeding, hybridization, and cultivation techniques paved the way for the diverse array of flower varieties that we cherish today.

Scope and Diversification of Floriculture

The scope of floriculture is vast and diversified, encompassing various facets within horticulture. From growing outdoor and indoor ornamental plants to producing cut flowers, potted plants, and landscape designs, floriculture offers a wide range of opportunities for horticulture enthusiasts. Specialized fields such as greenhouse management, floral design, flower breeding, and export-import trade provide avenues for skilled professionals. Additionally, floriculture intersects with other horticultural disciplines, such as plant propagation, plant physiology, and pest management, further enriching the horticulture syllabus.

Sustainable Practices in Floriculture

Incorporating sustainable practices in floriculture is vital for long-term environmental and economic viability. Employing integrated pest management (IPM) strategies to minimize the use of chemical pesticides, adopting eco-friendly fertilization methods, and implementing water-efficient irrigation techniques are essential components of sustainable floriculture. Cultivating native and climate-adapted plant species reduces the ecological footprint and conserves local

Flowers Cultivation in Horticulture

Flowers cultivation is a vital aspect of horticulture that focuses on the production and management of ornamental flowering plants. The significance of flowers cultivation and its role in enhancing the beauty and aesthetic value of landscapes, gardens, and urban spaces. The chapter explores the economic, cultural, and environmental importance of flowers and their contributions to the floriculture industry.

Selecting Suitable Flower Crops

Choosing the right flower crops for cultivation is crucial for successful flower production. This section provides insights into selecting flower species and varieties based on factors such as climate, soil type, market demand, and purpose (cut flowers, potted plants, landscape design). The chapter highlights popular flower crops and their characteristics, including roses, lilies, chrysanthemums, carnations, and tulips, among others.

Propagation Techniques for Flower Crops

Propagation is a crucial aspect of flower crop cultivation, enabling the multiplication and maintenance of desired genetic traits. Flower crops are propagated using various techniques, such as seeds, cuttings, division, grafting, and tissue culture. Seed propagation is common for annual flowers, while vegetative methods like cuttings and division are preferred for perennial varieties. Grafting allows the combination of desirable traits from different plants, while tissue culture ensures rapid mass multiplication of genetically identical plants. Selecting the appropriate propagation technique is essential to ensure successful flower cultivation, consistent bloom quality, and the preservation of unique varieties in the floral industry.

Site Selection and Planting Techniques

Choosing the right site and employing suitable planting techniques are essential for successful flower crop cultivation. Flower crops thrive in locations with ample sunlight, well-drained soil, and protection from strong winds. Proper site preparation, including soil testing and improvement, ensures optimal growing conditions. Planting techniques vary based on the flower type—seeds are sown directly for annuals, while transplants or bulbs are used for perennials. Correct spacing and planting depth are critical for healthy root development and vigorous growth. By carefully selecting the site and employing appropriate planting methods, flower growers can enhance bloom production, flower quality, and overall floral display.

Irrigation, Fertilization, and Pruning

Irrigation, fertilization, and pruning are essential practices in the successful cultivation of flower crops. Adequate irrigation ensures a consistent water supply, promoting healthy plant growth and vibrant blooms. Drip irrigation or soaker hoses are ideal for delivering water directly to the plant roots, minimizing water wastage. Appropriate fertilization with balanced nutrients encourages robust flowering and overall plant health. Organic fertilizers or slow-release formulations are often preferred to avoid excessive vegetative growth. Pruning is crucial for shaping and maintaining the desired form of flower plants, promoting better air circulation, and removing spent flowers to encourage continuous blooming. Proper implementation of these practices optimizes flower production, enhances the aesthetic appeal of the floral display, and extends the flowering period.

Pest and Disease Management

Effective pest and disease management is crucial to safeguard the health and beauty of flower crops. Integrated Pest Management (IPM) practices, which combine cultural, biological, and chemical control methods, are employed to minimize the impact of pests and diseases while ensuring environmental safety. Regular monitoring and early detection help identify potential issues, enabling timely interventions. Biological control, using natural predators and beneficial organisms, reduces reliance on chemical pesticides. Additionally, resistant varieties and proper sanitation practices are essential in preventing disease outbreaks. By adopting a proactive and holistic approach to pest and disease management, flower growers can maintain the quality and marketability of their floral products, ensuring customer satisfaction and sustainable floral production.

Types of Gardens in India

India is a land known for its rich horticultural heritage and diverse landscapes, which have given rise to various types of gardens across the country.

Mughal Gardens: Symmetry and Grandeur

Mughal gardens, inspired by Persian and Islamic designs, epitomize symmetry, grandeur, and meticulous planning. These gardens are characterized by geometric patterns, flowing water channels, fountains, and terraced landscapes. One of the most iconic Mughal gardens is the Taj Mahal's Charbagh, a four-part garden layout divided by water channels, representing paradise in Islamic traditions. Mughal gardens are famous for their chahar-bagh style, where the central axis leads to a pavilion or water feature, creating a mesmerizing visual effect.

Colonial Gardens: Timeless Elegance

Colonial gardens in India exhibit a blend of European and Indian design elements. During the British colonial period, British gardeners introduced European plants and landscaping techniques, adapting them to the Indian climate and culture. Colonial gardens often feature well-manicured lawns, flower beds, and avenues of trees. The Victoria Memorial Garden in Kolkata and the Government Botanical Garden in Ooty are exemplary instances of colonial garden design, combining British aesthetics with local flora and fauna.

Rock Gardens: Embracing the Terrain

Rock gardens in India are captivating landscapes that creatively use rocks, boulders, and native plants to mimic natural mountain environments. These gardens are particularly popular in hilly regions like the Himalayas. The Chandigarh Rock Garden, created by artist Nek Chand, is a famous example, showcasing an extensive collection of sculptures and artwork made from recycled materials. Rock gardens require thoughtful placement of rocks, creating crevices and niches for plants to thrive, and offering visitors a glimpse of rugged mountain terrain.

Terrace Gardens: Green Oases in Urban Spaces

Terrace gardens, also known as rooftop gardens, are increasingly popular in urban areas, especially in cities with limited green spaces. These gardens transform rooftops and balconies into lush green oases, providing a peaceful retreat amidst the concrete jungle. Terrace gardens are designed with lightweight containers and appropriate soil mixtures to accommodate the weight restrictions of the structures. They offer opportunities to grow a variety of plants, from ornamental flowers to vegetables and herbs, making them both aesthetically pleasing and functional.

Zen Gardens: Serenity and Meditation

Zen gardens, influenced by Japanese Zen Buddhism, are designed to evoke serenity, simplicity, and contemplation. These gardens often feature raked gravel or sand, carefully placed rocks, and pruned trees and shrubs. The Ryoanji Temple Garden in Kyoto, Japan, is an iconic Zen garden

with 15 rocks artfully arranged in a sea of raked gravel, inviting visitors to meditate and find inner peace. In India, Zen gardens are increasingly popular in meditation centers and wellness retreats, offering a tranquil escape from the hustle and bustle of everyday life.

Fruit and Vegetable Gardens: Sustainable Bounty

Fruit and vegetable gardens, commonly found in rural and suburban areas, focus on sustainable agriculture and self-sufficiency. These gardens incorporate a variety of fruit-bearing trees, bushes, and vines, along with vegetable crops. The concept of permaculture is often applied to ensure ecological balance and efficient use of resources. Fruit and vegetable gardens in India reflect the diversity of the country's climate and culture, showcasing regional favorites like mangoes, bananas, coconuts, and a plethora of vegetables.

Bonsai Gardens: Miniature Artistry

Bonsai gardens are an artistic expression of horticulture, where miniature trees are cultivated and pruned to resemble their full-sized counterparts in nature. The art of bonsai involves careful training and shaping of the tree's branches and roots to create aesthetically pleasing forms. Bonsai gardens are an exquisite display of patience and skill, showcasing the ingenuity of horticulturists in creating living masterpieces. Many botanical gardens in India boast impressive bonsai collections, offering visitors a glimpse into this captivating art form.

Harvesting, Post-Harvest Handling, and Marketing

Proper harvesting techniques for different flower crops to ensure peak quality and vase life. It delves into post-harvest handling practices, such as grading, sorting, and packaging, to maintain freshness and extend shelf life. The chapter also explores marketing channels, including local markets, florists, wholesale distribution, and online platforms, and the significance of meeting consumer demands and preferences.

In conclusion, the cultivation of flowers in horticulture is a multifaceted and rewarding endeavor. Understanding the unique requirements of flower crops, from their growth physiology to post-harvest handling, is crucial for successful flower production. As a critical component of the floriculture industry, flowers cultivation contributes to the enhancement of the environment's aesthetic appeal and the well-being of individuals, while offering economic opportunities for growers and floral enthusiasts alike.

Chapter 11

Olericulture: The Science of Vegetable Cultivation

Olericulture is a specialized branch of horticulture that focuses on the cultivation and management of vegetable crops.

Importance of Vegetable Crops

Vegetable crops are vital components of human diets, providing essential nutrients, vitamins, and minerals necessary for maintaining good health. Olericulture addresses the challenges of producing high-quality vegetables to meet the demands of both domestic and international markets. The importance of diverse vegetable crops in a balanced diet, their role in reducing malnutrition and promoting health, and the economic value of the vegetable industry.

Vegetable Classification and Morphology

This section delves into the classification of vegetable crops based on botanical families and their morphological characteristics. The different types of vegetables, such as leafy greens, root vegetables, crucifers, legumes, and solanaceous crops, helps horticulturists choose suitable cultivation practices and management techniques.

Propagation and Planting Techniques

Vegetable cultivation relies on various propagation and planting techniques to ensure successful crop establishment and optimal yields. Each technique has its advantages and is suitable for specific vegetable species. Here are some important notes on different propagation and planting techniques used in vegetable cultivation:

Seed Sowing:

- Seed sowing is the most common and economical method of vegetable propagation.
- It involves planting seeds directly into the soil or in seed trays/nurseries for later transplanting.
- Suitable for crops like tomatoes, peppers, cucumbers, and beans.
- Seed sowing allows for better control over seedling development and spacing.

Transplanting:

- Transplanting involves raising seedlings in nurseries and then moving them to the main field.

- This technique provides an advantage in regions with short growing seasons and adverse weather conditions.
- Commonly used for vegetables like tomatoes, peppers, cabbage, and broccoli.
- Transplanting ensures uniform plant growth and reduces competition among seedlings.

Grafting:

- Grafting is a specialized technique used to combine desirable traits from different plant varieties.
- It involves joining the stem of one plant (scion) onto the root system of another (rootstock).
- Grafting improves disease resistance, tolerance to environmental stress, and overall plant vigor.
- Commonly used for tomatoes, cucumbers, and melons to enhance productivity and disease resistance.

Vegetative Propagation:

- Vegetative propagation involves using plant parts other than seeds to grow new plants.
- Techniques like cuttings, layering, and division are used for this purpose.
- Suitable for crops like potatoes (through tubers), garlic (cloves), and mint (root divisions).
- Vegetative propagation helps maintain the genetic characteristics of desirable varieties.

Tissue Culture:

- Tissue culture is a modern technique used for rapid mass multiplication of genetically identical plants.
- It involves culturing plant cells, tissues, or organs in a sterile medium under controlled conditions.
- Tissue culture allows for disease-free plant production and is used for high-value crops like bananas and orchids.
- It enables the production of a large number of plants in a short time.

Direct Seeding and Broadcasting:

- Direct seeding involves sowing seeds directly into the field without raising seedlings in nurseries.
- Broadcasting refers to scattering seeds evenly across the field surface.
- Suitable for crops like radishes, carrots, and lettuce.
- Direct seeding and broadcasting are less labor-intensive but may require thinning to achieve proper plant spacing.

Each propagation and planting technique offers its own set of advantages and considerations. Successful vegetable cultivation relies on choosing the most suitable technique based on crop characteristics, growing conditions, and farmer preferences. Proper implementation of these techniques ensures healthy plant establishment, improved crop productivity, and ultimately contributes to food security and sustainable agriculture practices.

Soil and Nutrient Management

Soil Testing:

- Soil testing is the first step in vegetable soil and nutrient management.
- It involves analyzing soil samples to determine pH, nutrient levels, and other essential properties.
- Soil testing helps farmers understand the nutrient status of the soil and guides them in making informed decisions about fertilization.

Organic Matter and Soil Fertility:

- Organic matter is crucial for maintaining soil fertility and structure.
- Adding organic matter through composting, green manure, or crop residues enhances soil health and nutrient retention.
- Organic matter improves water-holding capacity, aeration, and microbial activity, benefiting vegetable growth.

Balanced Fertilization:

- Providing the right balance of essential nutrients is vital for vegetable crops.
- Balanced fertilization involves supplying macro and micronutrients in the correct proportion.
- Nitrogen, phosphorus, and potassium (NPK) are primary macronutrients, while elements like iron, zinc, and boron are important micronutrients.

Organic vs. Inorganic Fertilizers:

- Vegetable farmers can choose between organic and inorganic (chemical) fertilizers.
- Organic fertilizers, like compost and manure, release nutrients slowly and improve soil structure over time.
- Inorganic fertilizers provide immediate nutrient availability, but overuse can lead to soil degradation and environmental issues.

Nutrient Timing:

- Providing nutrients at the right time is critical for maximizing vegetable yields.
- Pre-plant fertilization ensures essential nutrients are available during early growth stages.
- Side-dressing or top-dressing with nitrogen during the growing season replenishes nutrient demands.

Nutrient Uptake and Efficiency:

- Understanding nutrient uptake patterns of different vegetable crops is essential.
- Crop-specific nutrient requirements help avoid over-fertilization and minimize nutrient losses.
- Efficient nutrient management reduces production costs and minimizes environmental impact.

Crop Rotation and Intercropping:

- Crop rotation and intercropping practices contribute to nutrient management.
- Rotating vegetable crops helps break disease cycles and reduces nutrient depletion.
- Intercropping enhances nutrient use efficiency, as different crops have varying nutrient requirements.

Water Management:

- Proper water management is linked to nutrient availability and uptake by vegetable plants.
- Over-irrigation can lead to nutrient leaching, while under-irrigation affects nutrient absorption.
- Implementing efficient irrigation techniques like drip irrigation helps conserve water and optimize nutrient delivery.

Cover Cropping:

- Cover crops, such as legumes and grasses, are planted between vegetable growing seasons.
- Cover crops improve soil fertility by fixing nitrogen and adding organic matter when incorporated into the soil.
- They also suppress weeds and protect the soil from erosion.

Effective soil and nutrient management practices ensure sustainable vegetable cultivation, increased productivity, and improved soil health. By adopting balanced fertilization, using organic matter, and optimizing water and nutrient delivery, farmers can achieve healthy and nutritious vegetable crops while preserving the long-term fertility and productivity of their land.

Traditional Open Field Cultivation

Traditional open field cultivation is the most common method of vegetable farming in India. This section covers the cultivation of vegetables in open fields using conventional agricultural practices. Farmers grow a wide range of vegetables, such as tomatoes, potatoes, onions, leafy greens, and cucurbits, using natural sunlight and rainwater.

Protected Cultivation: Greenhouses and Polyhouses

Protected cultivation has gained popularity in India, especially in regions with extreme weather conditions or limited arable land. The use of greenhouses and polyhouses to create controlled environments for vegetable crops. These structures offer protection against adverse weather, pests, and diseases, allowing year-round cultivation of high-value crops.

Urban and Peri-Urban Vegetable Farming

With rapid urbanization, urban and peri-urban vegetable farming has emerged as a sustainable solution to meet the demand for fresh produce in cities, the cultivation of vegetables in backyards, rooftop gardens, and community spaces within urban areas. The use of organic farming practices and composting to enhance soil fertility and minimize environmental impact. Urban agriculture also fosters a sense of community involvement and promotes access to nutritious vegetables in urban food deserts.

Vegetable Vertical Farming

Vertical farming is an innovative agricultural technique that involves cultivating vegetables in vertically stacked layers or shelves, often in controlled indoor environments. This approach optimizes land use, making it particularly suitable for urban areas with limited space. Vertical farming utilizes artificial lighting, efficient water-recirculating systems, and precise nutrient delivery to promote year-round vegetable production. By harnessing advanced technologies and automation, vertical farming maximizes resource efficiency, reduces water consumption, and minimizes pesticide usage, resulting in sustainable, high-yield vegetable cultivation.

Vegetable Hydroponics

Hydroponics is a soilless cultivation technique in which vegetable plants are grown directly in nutrient-rich water solutions, eliminating the need for traditional soil. This method allows for precise control over nutrient levels, pH, and environmental factors, promoting rapid plant growth and increased productivity. Hydroponics significantly reduces water usage compared to conventional farming and is ideal for regions with water scarcity. The absence of soil-borne diseases and weeds results in cleaner and healthier produce. Hydroponic vegetable cultivation is gaining popularity worldwide as a sustainable and resource-efficient approach to meet the growing demand for fresh vegetables.

Organic Vegetable Farming

Organic vegetable farming is gaining popularity in India due to the increasing demand for chemical-free produce. The principles and practices of organic farming, including the use of natural fertilizers, biopesticides, and crop rotation. It emphasizes the importance of conserving soil health and biodiversity while promoting sustainable agricultural practices. Organic vegetable cultivation supports farmers' economic well-being by commanding premium prices in both domestic and international markets.

Terrace and Balcony Gardening

Terrace and balcony gardening are creative and space-efficient techniques of growing vegetables in urban environments. These methods transform small outdoor spaces into productive green havens, allowing urban dwellers to grow their own fresh and nutritious produce. Using containers, pots, and vertical structures, a wide range of vegetables can be cultivated, including tomatoes, peppers, lettuce, herbs, and more. Vegetable terrace and balcony gardens not only enhance the aesthetics of urban spaces but also contribute to improved food security, reduced food miles, and a sense of self-sufficiency. By adopting organic practices and utilizing recycled materials, these gardens promote sustainable living and reconnect city dwellers with the joys of nurturing and harvesting their homegrown vegetables.

Irrigation and Water Management

Efficient irrigation and water management are vital for successful vegetable cultivation. Proper watering ensures consistent and adequate moisture supply, supporting optimal plant growth and yield. Drip irrigation, sprinkler systems, and rainwater harvesting are popular methods used to conserve water and minimize wastage. Monitoring soil moisture levels and adjusting irrigation schedules based on crop needs are essential practices. Mulching is another effective technique that reduces evaporation, conserves soil moisture, and suppresses weed growth. By adopting water-efficient irrigation methods and implementing smart water management strategies, vegetable growers can enhance crop productivity, reduce water usage, and contribute to sustainable agricultural practices.

Integrated Pest and Disease Management

Integrated Pest and Disease Management (IPM) is a holistic approach to control pests and diseases in vegetable crops while minimizing environmental and health risks. IPM combines various strategies, including cultural practices, biological control, physical barriers, and judicious use of pesticides, to prevent, monitor, and manage pests and diseases. Crop rotation, intercropping, and maintaining proper plant spacing reduce pest populations, while beneficial insects and natural enemies help control pest outbreaks. Regular monitoring and early detection enable timely interventions, reducing the need for chemical pesticides. By promoting a balanced ecosystem and employing IPM techniques, vegetable growers can achieve sustainable pest and disease control, ensuring healthy and high-quality produce.

Harvesting and Post-Harvest Handling

Harvesting and post-harvest handling are crucial stages in vegetable production that directly impact the quality and shelf life of the produce. Timely harvesting ensures vegetables are picked at their peak ripeness, preserving nutritional content and flavor. Proper handling, sorting, and grading techniques minimize physical damage and reduce post-harvest losses. Rapid cooling and storage in controlled environments maintain freshness and extend shelf life. Post-harvest treatments like washing, sanitizing, and packaging help maintain hygiene and food safety. Implementing efficient post-harvest practices ensures that consumers have access to fresh, nutritious, and safe vegetables, while also maximizing profits for farmers and stakeholders along the supply chain.

Conclusion

The diverse types of vegetable cultivation in India demonstrate the country's adaptability and innovation in agricultural practices. From traditional open field cultivation to modern protected cultivation techniques, each method offers unique opportunities and challenges. As India strives to enhance food security and sustainability, the incorporation of diverse vegetable cultivation practices is crucial. By embracing innovative technologies and sustainable approaches, Indian farmers can optimize vegetable productivity and contribute to the nation's nutritional well-being. Additionally, supporting urban and peri-urban vegetable farming can address the growing demand for fresh produce in urban areas and foster community engagement in agriculture. It is essential for policymakers, researchers, and farmers to collaborate in promoting a resilient and diverse vegetable cultivation landscape for a healthy and prosperous India.

Chapter 12

Plant Breeding and Genetics

Principles of Plant Breeding

Plant breeding is a systematic and scientific process that aims to develop new and improved plant varieties with desirable traits. It involves the controlled crossing of plants to transfer specific genes and characteristics from one plant to another. Plant breeding plays a crucial role in agriculture and horticulture by creating crops with increased yield, disease resistance, nutritional value, and adaptability to changing environmental conditions. In this section, we will explore the fundamental principles of plant breeding, including the different breeding methods and considerations for successful breeding programs.

Objectives of Plant Breeding:

The primary objectives of plant breeding are to:

1. Improve Yield: Develop varieties with higher yield potential to meet the increasing demands for food and other plant-derived products.
2. Enhance Quality: Improve the quality of crops by enhancing traits such as flavor, nutritional content, and post-harvest characteristics.
3. Disease Resistance: Develop plant varieties with increased resistance to pests and diseases to reduce the reliance on chemical pesticides.
4. Abiotic Stress Tolerance: Create plants with better tolerance to environmental stresses like drought, heat, and salinity, ensuring stable yields under adverse conditions.
5. Adaptation: Breed crops that are well-adapted to different agro-climatic regions and specific growing conditions.

Breeding Methods:

Plant breeding employs several methods to create new plant varieties. The main breeding methods include:

1. Hybridization: Controlled crossing of two genetically diverse plants (usually referred to as parents) to produce hybrid offspring with desirable traits.
2. Selection: Choosing individual plants with desirable characteristics from a population and propagating them to create improved lines.
3. Mutagenesis: Inducing genetic mutations using chemicals or radiation to create genetic variability for selection of improved traits.
4. Biotechnology and Genetic Engineering: Introducing specific genes from one organism into another to confer desired traits, such as insect resistance or improved nutritional content.

Considerations for Successful Plant Breeding Programs:

Successful plant breeding programs require careful planning and implementation. Considerations for effective plant breeding include:

1. Genetic Diversity: Accessing a diverse gene pool is crucial to introduce new traits and improve crop resilience.
2. Breeding Objectives: Clearly define breeding objectives to guide the selection of parental lines and traits to be improved.
3. Parental Selection: Choose appropriate parents with complementary traits to maximize the chances of producing desired offspring.
4. Evaluation and Selection: Systematically evaluate and select promising plants with improved traits in multiple generations.
5. Replication and Randomization: Use replication and randomization in field trials to ensure accurate evaluation of breeding lines.
6. Multi-Location Testing: Conduct trials in various locations to assess the stability of the new variety across different environments.
7. Participatory Plant Breeding: Involve farmers and end-users in the breeding process to ensure the development of varieties that meet their specific needs.

Intellectual Property and Regulation:

Plant breeding often involves the creation of new plant varieties, leading to issues of intellectual property and regulation. Plant breeders' rights and patents are granted to protect the rights of breeders and incentivize continued investment in breeding research and development.

Plant breeding is a dynamic and ever-evolving field that significantly contributes to global food security and agricultural sustainability. By adhering to the principles of plant breeding and utilizing diverse breeding methods, plant breeders can create improved crop varieties that meet the growing challenges of the agricultural industry and address the needs of a changing world.

Traditional Breeding Methods

Traditional breeding methods are time-tested techniques that have been used for centuries to improve plants' genetic traits and develop new crop varieties. These methods involve controlled crossing of plants with desirable traits to create offspring with improved characteristics. Traditional breeding plays a significant role in plant breeding and has contributed to the development of many of today's cultivated crop varieties. In this section, we will explore the principles and various traditional breeding methods used in plant breeding.

Principles of Traditional Breeding:

Traditional breeding methods are based on the principles of heredity and genetics, aiming to combine desirable traits from different parent plants to create offspring with improved characteristics. The key principles include:

1. Genetic Variation: Genetic variation within a species provides the raw material for breeding new varieties with different traits.
2. Controlled Cross-Pollination: Controlled pollination involves transferring pollen from one selected parent plant to another to ensure specific traits are passed on to the offspring.
3. Selection: After cross-pollination, the best plants with desired traits are selected to be the parents of the next generation, leading to a gradual accumulation of favorable traits over time.
4. Repetition and Successive Generations: The breeding process is repeated over multiple generations to stabilize and enhance the desired traits in the new variety.

Traditional Breeding Methods:

Several traditional breeding methods are used in plant breeding. The main methods include:

1. Inbreeding: Inbreeding involves crossing closely related plants to achieve uniformity and fix desirable traits in the offspring. This method is used to develop pure lines or inbred lines.
2. Outcrossing: Outcrossing involves crossing unrelated plants to introduce genetic diversity into the breeding program. This method enhances hybrid vigor and increases genetic variability.
3. Backcrossing: Backcrossing is used to transfer a specific trait from one plant (donor parent) to another (recipient parent). The offspring resulting from the cross are then repeatedly crossed back to the recipient parent until the trait of interest is fixed.
4. Mass Selection: Mass selection involves selecting and propagating seeds from the best-performing plants in a population. This method is useful for improving simple, quantitative traits.
5. Pure Line Selection: Pure line selection is a form of inbreeding that aims to develop homozygous lines with uniform traits.
6. Pedigree Breeding: Pedigree breeding involves keeping detailed records of the parentage of each plant in the breeding program. It is useful for tracking specific traits through multiple generations.
7. Bulk Breeding: Bulk breeding involves mixing seeds from multiple plants in a population and propagating them together. This method is often used for self-pollinating crops.

Advantages and Limitations of Traditional Breeding Methods:

Advantages of traditional breeding methods include:

- Cost-Effectiveness: Traditional breeding is relatively low-cost compared to biotechnological approaches.
- Wide Adoption: Traditional breeding methods are well-established and widely adopted by breeders around the world.
- Genetic Diversity: These methods allow for the preservation and enhancement of genetic diversity within crop populations.

Limitations of traditional breeding methods include:

- Time-Consuming: Traditional breeding methods require multiple generations to develop new varieties.
- Limited Precision: These methods may not offer the same precision as biotechnological approaches in introducing specific genes.

Modernization of Traditional Breeding:

While traditional breeding methods remain valuable, modern plant breeding often combines traditional techniques with biotechnological tools like marker-assisted selection (MAS) and genomic selection. These advancements enhance the efficiency and precision of traditional breeding, leading to the development of improved crop varieties with targeted traits.

Traditional breeding methods have played a critical role in the development of diverse, resilient, and high-yielding crop varieties that feed the world's population. By harnessing the principles of genetic variation, controlled crossing, and selection, traditional breeding continues to be an essential tool in addressing global challenges in agriculture and ensuring food security for future generations.

Biotechnological Approaches to Plant Breeding

Biotechnological approaches to horticultural plant breeding have revolutionized the field of plant genetics and breeding by offering powerful tools to manipulate plant genomes and introduce desirable traits more precisely and efficiently. These approaches involve the use of genetic engineering, molecular markers, and genomics to accelerate the breeding process and create improved crop varieties. In this section, we will explore the principles and various biotechnological approaches used in horticultural plant breeding.

Principles of Biotechnological Approaches:

Biotechnological approaches to horticultural plant breeding are based on the principles of molecular genetics and genetic manipulation. Key principles include:

1. Genetic Engineering: Genetic engineering involves the direct manipulation of an organism's DNA using recombinant DNA technology to introduce specific genes or modify existing ones.
2. Molecular Markers: Molecular markers are specific DNA sequences associated with desirable traits that can be used to track and select plants with those traits more efficiently.
3. Genomics: Genomics is the study of the entire DNA sequence of an organism, allowing for a comprehensive understanding of gene function and regulation.

Biotechnological Approaches in Horticultural Plant Breeding:

Several biotechnological approaches are used in horticultural plant breeding. The main approaches include:

1. Genetic Engineering (Transgenic Plants): Genetic engineering involves introducing genes from other organisms into horticultural crops to confer specific traits, such as insect resistance, disease resistance, or improved nutritional content.
2. Marker-Assisted Selection (MAS): MAS uses molecular markers linked to specific genes or traits to identify and select plants with desired traits more efficiently. This approach accelerates the breeding process by reducing the need for time-consuming and resource-intensive field evaluations.
3. Genomic Selection: Genomic selection uses high-throughput genotyping and phenotyping to predict the performance of plants based on their entire genome. This approach enables breeders to select superior individuals at early stages of development.
4. Gene Editing: Gene editing techniques, such as CRISPR-Cas9, allow for precise modifications of specific genes within a plant's genome. This approach has the potential to create targeted changes to improve crop traits without introducing foreign DNA.
5. RNA Interference (RNAi): RNAi is a process that silences specific genes by introducing double-stranded RNA molecules that target and degrade the corresponding messenger RNA. This approach can be used to downregulate genes associated with undesirable traits.

Advantages of biotechnological approaches to horticultural plant breeding include:

- Precision: These approaches allow for precise manipulation of genes to introduce specific traits.
- Speed: Biotechnological approaches can expedite the breeding process, reducing the time required to develop improved varieties.
- Efficiency: Molecular markers and genomics enable more efficient selection of desired traits.

Limitations of biotechnological approaches include:

- Regulatory Constraints: Genetically modified organisms (GMOs) may face regulatory hurdles in

some regions, limiting their commercial adoption.
- Public Perception: Public perception and acceptance of GMOs can influence their marketability.

Ethical and Regulatory Considerations:

Biotechnological approaches to horticultural plant breeding raise ethical and regulatory considerations related to environmental impact, food safety, and intellectual property rights. Responsible use of these technologies requires adherence to ethical guidelines, rigorous safety assessments, and transparent communication with stakeholders.

Biotechnological approaches to horticultural plant breeding have significantly advanced our ability to develop improved crop varieties with desired traits. By leveraging the principles of genetic engineering, molecular markers, and genomics, plant breeders can address challenges in horticulture more effectively, contributing to sustainable agriculture and the continued progress of the horticultural industry.

Chapter 13:

Greenhouse Horticulture

Greenhouse horticulture is a specialized method of crop production that takes place within controlled environments, typically enclosed structures made of transparent materials like glass or plastic. Greenhouses offer numerous advantages, such as extended growing seasons, protection from adverse weather conditions, and enhanced crop quality. In this section, we will explore the different types of greenhouses and their key components, which play a crucial role in creating and maintaining the optimal growing environment.

Types of Greenhouses:

1. Traditional Glass Greenhouses: Traditional glass greenhouses are the classic structures made of glass panels. They provide excellent light transmission and create a stable environment for plant growth. They are commonly used for commercial production of various horticultural crops.
2. Plastic-Film Greenhouses: Plastic-film greenhouses are made of polyethylene or other plastic materials. They are more cost-effective than glass greenhouses and are often used for seasonal or temporary crop production.
3. Shade Houses: Shade houses are greenhouses covered with shade cloth that reduces the amount of sunlight reaching the crops. They are used for shade-loving plants or to protect sensitive plants from intense sunlight.
4. Multi-Span Greenhouses: Multi-span greenhouses are large structures made up of multiple connected bays. They offer efficient space utilization and are widely used for commercial-scale crop production.
5. High-Tunnel Greenhouses: High-tunnel greenhouses are unheated, low-cost structures with high arches, usually made of plastic film. They are used to extend the growing season in temperate regions and protect crops from frost.
6. Lean-To Greenhouses: Lean-to greenhouses are attached to existing buildings, utilizing one wall for support. They are suitable for limited space and are often used for home gardening or research purposes.

Key Components of Greenhouses:

1. Frame: The frame provides the structural support for the greenhouse. It can be made of materials like steel, aluminum, or wood, depending on the greenhouse type and size.
2. Covering Material: The covering material determines the amount of light transmission and insulation in the greenhouse. Common materials include glass, polyethylene, polycarbonate, or acrylic.
3. Ventilation System: A ventilation system is crucial for regulating temperature and humidity within the greenhouse. It typically includes roof vents, side vents, or mechanical fans.
4. Heating System: In regions with cold climates, a heating system is essential to maintain a suitable temperature for plant growth during colder months.
5. Cooling System: In warmer climates or during hot periods, a cooling system is needed to prevent

overheating. Options include shade cloths, evaporative cooling, or fans.

6. Irrigation System: An efficient irrigation system ensures plants receive an adequate and consistent water supply. Options include drip irrigation, sprinklers, or ebb and flow systems.
7. Benches and Shelves: Benches and shelves provide elevated platforms for plant placement, optimizing space usage and facilitating plant care.
8. Environmental Control Systems: Modern greenhouses often incorporate automated environmental control systems that monitor and adjust temperature, humidity, and light levels to create optimal growing conditions.
9. Supplementary Lighting: In regions with limited natural light, supplementary lighting systems provide artificial light to extend the day length and promote plant growth.
10. Environmental Sensors: Environmental sensors, such as temperature and humidity sensors, help monitor the greenhouse conditions and inform the control systems.
11. Shade Systems: Shade systems, like retractable shade cloths, are used to control light intensity and protect crops from excessive heat and sunlight.

Greenhouses are versatile structures that allow horticulturists to grow a wide range of crops in controlled environments. The choice of greenhouse type and components depends on the specific needs of the crops, the climate, and the intended scale of production. By providing a favorable environment for plant growth, greenhouses have become essential tools in modern horticulture, contributing to increased productivity and consistent crop quality throughout the year.

Greenhouse Management: Temperature, Humidity, and Ventilation

Effective management of temperature, humidity, and ventilation is crucial for creating the optimal growing environment within a greenhouse. These factors significantly influence plant growth, development, and overall crop health. Proper greenhouse management ensures that horticultural crops thrive, leading to increased yields and improved crop quality. In this section, we will explore the key aspects of greenhouse management related to temperature, humidity, and ventilation.

Temperature Management:

Temperature is one of the most critical environmental factors affecting plant growth in a greenhouse. Different crops have specific temperature requirements, and maintaining the right temperature range is essential for maximizing their productivity and quality. Greenhouse temperature management involves the following aspects:

1. Heating Systems: Greenhouses located in cooler climates require heating systems to maintain the desired temperature during cold periods. Common heating methods include hot water or steam heating, forced-air heating, and radiant heating.
2. Cooling Systems: In warmer climates or during hot periods, cooling systems are essential to prevent excessive heat stress on plants. Cooling options include natural ventilation, mechanical ventilation, evaporative cooling, and shade cloths.
3. Temperature Setpoints: Greenhouse managers should establish temperature setpoints based on the crop's growth stage, external weather conditions, and time of day. Automated environmental

control systems can help maintain the desired temperature range.

4. Temperature Monitoring: Regularly monitor greenhouse temperatures using thermometers or temperature sensors to ensure that conditions remain within the target range.

5. Thermal Screens: Thermal screens or energy curtains can be used at night to retain heat within the greenhouse and improve energy efficiency.

Humidity Management:

Humidity levels in the greenhouse directly impact transpiration rates, water uptake, and disease incidence in plants. Proper humidity management is crucial for optimal plant growth. Key considerations for humidity management include:

1. Humidity Sensors: Install humidity sensors to monitor and control humidity levels accurately.

2. Humidification Systems: Humidification systems can be used to increase humidity levels when needed, especially during dry periods.

3. Ventilation: Proper ventilation helps regulate humidity levels by allowing excess moisture to escape from the greenhouse.

4. Avoid Overwatering: Overwatering can lead to high humidity levels and increase the risk of diseases, so it is essential to implement efficient irrigation practices.

Ventilation Management:

Ventilation is vital for maintaining proper airflow within the greenhouse. Adequate ventilation serves several purposes:

1. Temperature Regulation: Ventilation helps prevent overheating by allowing hot air to escape from the greenhouse and facilitating the entry of cooler air.

2. Humidity Control: Proper ventilation helps reduce humidity levels by allowing moisture-laden air to escape.

3. Disease Prevention: Good airflow helps minimize humidity and prevents the buildup of disease-causing pathogens.

4. Carbon Dioxide Management: Ventilation ensures an adequate supply of carbon dioxide for photosynthesis, promoting plant growth.

Key practices for ventilation management include:

1. Roof Vents: Roof vents are used to release hot air and provide passive ventilation when temperatures rise.

2. Side Vents: Side vents are located along the sides of the greenhouse and facilitate air exchange.

3. Mechanical Fans: Mechanical fans can be used to improve airflow and provide additional ventilation, especially during still, hot conditions.
4. Natural Ventilation: Natural ventilation relies on the greenhouse's physical design and prevailing wind direction to drive airflow.

Proper greenhouse management of temperature, humidity, and ventilation creates a controlled and favorable environment for horticultural crops. By ensuring the right conditions for growth and development, greenhouse managers can optimize crop productivity, reduce pest and disease pressure, and produce high-quality crops year-round. Efficient use of technology and monitoring systems enhances greenhouse management, allowing horticulturists to respond quickly to changing environmental conditions and achieve the best possible outcomes for their crops.

Hydroponics and Soilless Culture

Hydroponics and soilless culture are innovative methods of crop production that involve growing plants without traditional soil. Instead, plants receive essential nutrients through water-based solutions, providing precise control over nutrient delivery and optimizing resource use. Hydroponics and soilless culture are popular techniques in greenhouse horticulture, as they offer several advantages, including increased crop yields, reduced water usage, and minimized disease pressure. In this section, we will explore the principles, types, and benefits of hydroponics and soilless culture in greenhouse horticulture.

Principles of Hydroponics and Soilless Culture:

The primary principles of hydroponics and soilless culture involve providing plants with essential nutrients dissolved in water, maintaining proper pH levels, and ensuring adequate aeration of the root zone. In these systems, plants rely on a growing medium to support their roots while receiving a nutrient-rich solution directly. The key principles include:

1. Nutrient Solution: Plants in hydroponic systems receive a balanced nutrient solution containing essential mineral elements required for their growth and development.
2. Growing Medium: A growing medium, such as perlite, rockwool, coconut coir, or hydroton, supports the plants' roots while allowing efficient nutrient uptake.
3. pH Regulation: Maintaining the appropriate pH level of the nutrient solution is critical for optimal nutrient availability to plants.
4. Aeration: Ensuring adequate aeration of the root zone prevents root suffocation and enhances nutrient uptake.

Types of Hydroponic Systems:

Several types of hydroponic systems are used in greenhouse horticulture, each offering unique advantages. Common hydroponic systems include:

1. Nutrient Film Technique (NFT): In the NFT system, a thin film of nutrient solution continuously flows over the roots of plants, providing them with nutrients. The excess solution is then

collected and recirculated.

2. Deep Water Culture (DWC): DWC systems involve suspending plant roots in a nutrient solution. Air stones or diffusers provide aeration to the solution, ensuring sufficient oxygen for the roots.
3. Ebb and Flow (Flood and Drain): In an ebb and flow system, nutrient solution periodically floods the growing medium and is then drained back into a reservoir. This cycle provides aeration to the roots.
4. Drip System: Drip systems deliver nutrient solutions directly to the base of each plant through drip emitters or tubing, providing precise nutrient delivery.
5. Aeroponics: In aeroponics, plant roots are suspended in the air, and nutrient solution is misted or sprayed directly onto the roots, ensuring optimal aeration and nutrient uptake.

Benefits of Hydroponics and Soilless Culture:

Hydroponics and soilless culture offer numerous advantages in greenhouse horticulture:

1. Increased Crop Yields: Nutrients are readily available to plants, leading to faster growth rates and higher yields.
2. Water Efficiency: Hydroponic systems use significantly less water compared to traditional soil-based agriculture.
3. Disease Control: Soilless culture reduces the risk of soil-borne diseases, enhancing crop health.
4. Controlled Environment: Hydroponic systems in greenhouses allow for precise control over environmental conditions, optimizing plant growth.
5. Space Optimization: Hydroponic systems can be designed vertically, making efficient use of space in greenhouses.

Considerations for Successful Hydroponic and Soilless Culture:
To ensure successful hydroponic and soilless culture in greenhouse horticulture, several key considerations should be taken into account:

1. Nutrient Management: Regular monitoring and adjustment of the nutrient solution are crucial to provide plants with the correct nutrient balance.
2. pH Monitoring: Proper pH levels are vital for nutrient uptake. Regular monitoring and pH adjustments are necessary to maintain optimal conditions.
3. Aeration: Ensuring adequate aeration of the root zone prevents root rot and promotes healthy plant growth.
4. Sterilization: Growing media and equipment should be sterilized to prevent the spread of diseases and pathogens.
5. Temperature and Humidity Control: Greenhouse environmental control is essential to maintaining the ideal growing conditions for hydroponic systems.

Hydroponics and soilless culture have transformed greenhouse horticulture by offering a sustainable and efficient method of crop production. By eliminating the reliance on soil and optimizing nutrient delivery, these systems maximize resource use and enable year-round crop production in controlled environments. As horticulturists continue to embrace these advanced techniques, hydroponics and soilless culture will play an increasingly significant role in meeting the global demand for high-quality, fresh produce.

Chapter 14

Urban and Vertical Horticulture

Urban horticulture refers to the practice of growing and cultivating plants in cities and other urban areas. It plays a vital role in enhancing the quality of life for urban dwellers, promoting environmental sustainability, and addressing various challenges faced by cities. Horticulture in urban environments has gained increasing importance due to the rapid urbanization and the need to create greener, healthier, and more sustainable cities. In this section, we will explore the significance of horticulture in urban areas and its multifaceted benefits.

Enhancing Aesthetics and Green Spaces:

Urban horticulture adds natural beauty and enhances the aesthetics of urban landscapes. Parks, gardens, street plantings, and rooftop gardens provide lush green spaces that serve as visual and sensory oases amid concrete and steel. These green areas contribute to a sense of well-being, reduce stress, and provide opportunities for relaxation and recreation.

Improving Air Quality and Mitigating Urban Heat Island Effect:

Plants in urban environments play a crucial role in mitigating air pollution and improving air quality. They act as natural air filters, absorbing harmful pollutants such as carbon dioxide, nitrogen oxides, and particulate matter. By reducing air pollution, horticulture helps in combating respiratory issues and promoting better overall health for urban residents. Additionally, green spaces contribute to reducing the urban heat island effect by providing shade and cooling the environment through transpiration.

Storm water Management:

In cities with extensive impervious surfaces, such as roads and buildings, stormwater runoff can lead to flooding and water pollution. Urban horticulture, particularly the use of green roofs and rain gardens, helps manage stormwater by absorbing and storing rainwater. These green features allow water to be gradually released back into the environment, reducing the burden on urban drainage systems and preventing pollution of natural water bodies.

Urban Agriculture and Food Security:

Urban horticulture plays a vital role in urban agriculture, where fruits, vegetables, and herbs are grown in small-scale urban spaces. Urban farming promotes local food production, reduces food transportation costs, and improves food security by providing fresh, nutritious produce to urban communities. Community gardens and urban farms also serve as educational and social hubs, fostering community cohesion and environmental awareness.

Biodiversity Conservation:

By creating green spaces and diverse habitats, urban horticulture contributes to biodiversity conservation within cities. Green roofs, urban gardens, and naturalized areas support a wide range of plant and animal species, helping to maintain urban biodiversity and ecological balance.

Climate Change Mitigation and Adaptation:

Urban horticulture plays a role in climate change mitigation by sequestering carbon dioxide and reducing greenhouse gas emissions through improved air quality and energy conservation. It also contributes to climate change adaptation by creating resilient urban landscapes that can withstand the challenges of extreme weather events and changing climatic conditions.

Economic Benefits:

Horticulture in urban environments offers economic benefits through increased property values, tourism, and job creation. Well-maintained green spaces enhance the appeal of urban neighborhoods, attracting businesses, residents, and visitors alike.

Horticulture in urban environments is essential for creating livable, sustainable, and resilient cities. Its multifaceted benefits extend to environmental, social, and economic domains, making urban horticulture a vital component of urban planning and development strategies. By integrating horticulture into urban landscapes, cities can thrive as vibrant, green, and healthy environments that improve the well-being and quality of life for their inhabitants.

Importance of Horticulture in Urban Environments: Rooftop Gardening, Balcony Gardens, and Terrace Farming, Vertical Gardening, and Living Walls

In recent years, the rapid urbanization of cities has led to a significant decrease in green spaces and natural landscapes. This phenomenon has raised concerns about environmental degradation, air and water pollution, and the loss of biodiversity. Horticulture, the practice of growing and cultivating plants, plays a crucial role in transforming urban environments into greener, healthier, and more sustainable spaces. In this chapter, we will explore the importance of horticulture in urban areas, with a focus on rooftop gardening, balcony gardens, terrace farming, vertical gardening, and living walls.

Importance of Horticulture in Urban Environments:

Urban horticulture is of paramount importance for various reasons:

Environmental Benefits:

Horticultural practices in urban environments contribute to environmental conservation and restoration. Green spaces created through rooftop gardening, balcony gardens, and living walls act as natural air purifiers, absorbing harmful pollutants and carbon dioxide. These green areas help mitigate the urban heat island effect, reducing the overall temperature of cities and conserving energy.

Biodiversity Conservation:

Urban horticulture promotes biodiversity conservation within cities. By cultivating a variety of plant species in different settings, such as vertical gardens and terrace farms, horticulturists create habitats for insects, birds, and other wildlife. These micro-habitats play a critical role in maintaining urban biodiversity and ecological balance.

Food Security and Local Food Production:

Rooftop gardening, balcony gardens, and terrace farming contribute to urban agriculture and local food production. Growing vegetables, fruits, and herbs in small spaces helps address food security concerns by providing fresh and nutritious produce to urban communities. Additionally, urban agriculture reduces the carbon footprint associated with long-distance transportation of food.

Community Engagement and Social Cohesion:

Horticultural initiatives, such as community gardens and green spaces, foster community engagement and social cohesion. Residents come together to take care of shared gardens, building strong bonds and a sense of belonging. These spaces become centers for community events, educational programs, and recreational activities, enhancing the overall quality of life in urban areas.

Rooftop Gardening, Balcony Gardens, and Terrace Farming:

Rooftop gardening, balcony gardens, and terrace farming are innovative horticultural practices that make the most of limited urban spaces.

Rooftop Gardening:

Rooftop gardening involves transforming rooftops into green spaces by cultivating a variety of plants. With proper planning and engineering, rooftops can support gardens that range from small containers to extensive green roofs. Rooftop gardens offer numerous benefits, including stormwater management, insulation, and energy efficiency. They also provide recreational areas for residents, improving mental well-being and enhancing the visual appeal of buildings.

Balcony Gardens:

Balcony gardens are created by using containers and vertical planters to grow plants on balconies and terraces. These gardens are ideal for individuals with limited space but a passion for gardening. Balcony gardens offer a private outdoor oasis, providing a space for relaxation and enjoyment of nature within the confines of an urban setting.

Terrace Farming:

Terrace farming involves the cultivation of crops on terraced slopes or steps built into hilly or sloping areas. In urban environments, terrace farming is often practiced on rooftops or elevated surfaces. This method optimizes space and water use, allowing urban dwellers to grow a variety of crops, including vegetables, herbs, and even small fruit trees.

Vertical Gardening and Living Walls:

Vertical gardening and living walls are innovative approaches to incorporate greenery into urban spaces while maximizing the use of vertical surfaces.

Vertical Gardening:

Vertical gardening utilizes vertical structures, such as trellises, walls, and fences, to grow plants vertically. It is particularly useful for small spaces where horizontal gardening is limited. Vining plants, such as tomatoes, cucumbers, and climbing flowers, thrive in vertical gardens. This approach adds aesthetic value to urban spaces, creating a lush and visually appealing environment.

Living Walls (Green Walls):

Living walls, also known as green walls, are vertical structures covered with plants. These walls can be freestanding or attached to buildings. Living walls use a variety of techniques, including hydroponics and soilless culture, to support plant growth. They offer numerous benefits, such as air purification, thermal insulation, and noise reduction. Living walls also add biodiversity to urban environments, attracting insects and providing habitat for birds.

In conclusion, horticulture in urban environments plays a pivotal role in enhancing the quality of life for urban dwellers and promoting sustainability. Rooftop gardening, balcony gardens, terrace farming, vertical gardening, and living walls offer creative and practical solutions to the challenges posed by urbanization. By embracing horticulture in cities, we can create greener, healthier, and more vibrant urban spaces that benefit both people and the environment. These horticultural practices empower individuals and communities to actively participate in shaping the future of their cities and fostering a closer connection with nature amidst the urban jungle.

Chapter 15:

Sustainable Horticulture

Sustainable horticulture is a holistic approach to farming and gardening that aims to meet the needs of the present generation without compromising the ability of future generations to meet their own needs. It encompasses environmentally responsible practices, social equity, and economic viability. In this chapter, we will explore the principles of sustainable agriculture and delve into the principles and practices of organic horticulture.

Principles of Sustainable Agriculture

Biodiversity Conservation:

Sustainable horticulture emphasizes the preservation and enhancement of biodiversity. It involves cultivating a diverse range of crops, incorporating native plant species, and creating habitats for beneficial insects and wildlife. Biodiversity conservation helps maintain ecological balance, increases resilience to pests and diseases, and promotes overall ecosystem health.

Soil Health and Fertility:

Sustainable horticulture prioritizes soil health as the foundation of agricultural productivity. Practices such as cover cropping, crop rotation, and organic matter incorporation improve soil structure, enhance nutrient cycling, and increase soil fertility. Healthy soils support vigorous plant growth and reduce the need for synthetic fertilizers.

Water Conservation and Management:

Water is a precious resource, and sustainable horticulture adopts water-efficient practices. Techniques like drip irrigation, rainwater harvesting, and mulching reduce water wastage and promote efficient water use. Water management strategies help mitigate the impact of water scarcity and ensure the sustainability of horticultural practices.

Integrated Pest Management (IPM):

Sustainable horticulture embraces the principles of Integrated Pest Management (IPM) to manage pests, diseases, and weeds. IPM combines biological, cultural, physical, and chemical control methods to minimize pesticide use while maintaining pest populations at an acceptable level. This approach reduces the impact on beneficial organisms and the environment.

Energy Efficiency and Renewable Resources:

Sustainable horticulture seeks to reduce energy consumption and reliance on non-renewable resources. Utilizing renewable energy sources, such as solar power for irrigation and greenhouse operations, helps lower greenhouse gas emissions and promotes environmental sustainability.

Social Equity and Community Engagement:

Sustainable horticulture considers the social aspect of farming, ensuring fair wages, safe working conditions, and the promotion of social equity among agricultural workers. Community engagement and involvement in decision-making processes foster a sense of ownership and responsibility for sustainable practices.

Organic Horticulture: Principles and Practices:

Organic horticulture is an integral part of sustainable horticulture, focusing on the use of natural inputs and practices that minimize the use of synthetic chemicals and promote ecological balance.

Soil Management:

Organic horticulture emphasizes soil health through the use of organic matter, compost, and green manures. These practices enrich the soil with essential nutrients, improve water retention, and enhance soil biodiversity.

Crop Diversity:

Growing a diverse range of crops helps reduce the risk of pest and disease outbreaks, improves soil fertility, and ensures a balanced and nutritious diet. Crop rotation and companion planting are common practices in organic horticulture.

Biological Pest Control:

Organic horticulture relies on biological control methods to manage pests. Beneficial insects, such as ladybugs and predatory nematodes, are introduced to control pest populations naturally.

Non-Chemical Weed Management:

Organic horticulture employs various non-chemical weed management strategies, such as mulching, hand weeding, and mechanical cultivation. These practices prevent weed competition without relying on synthetic herbicides.

Prohibition of Synthetic Chemicals:

Organic horticulture strictly prohibits the use of synthetic pesticides, herbicides, and genetically modified organisms (GMOs). Instead, natural pest and disease control methods are preferred to maintain ecological balance.

Certification and Standards:

Organic horticulture adheres to specific certification standards, ensuring that the practices and products meet organic principles. Certifications provide transparency and consumer confidence in organic produce.

In conclusion, sustainable horticulture, with a focus on organic principles and practices, is essential for safeguarding the health of ecosystems, human well-being, and the future of agriculture. By embracing biodiversity conservation, prioritizing soil health, conserving water,employing integrated pest management, and promoting organic horticultural practices, we can create a resilient and sustainable food system. Sustainable and organic horticulture not only protects the environment but also supports local communities, enhances food quality, and fosters a more harmonious relationship between humans and nature.

Chapter 16:

Agroecological Approaches and Biodiversity Conservation of Crops

Agroecological approaches in horticulture prioritize the integration of ecological principles into agricultural practices. These methods aim to optimize crop productivity while promoting biodiversity conservation, enhancing ecosystem services, and ensuring the sustainability of agricultural systems. In this chapter, we will explore the key concepts and practices of agroecology and how they contribute to biodiversity conservation in horticultural crops.

Agroccological Principles in Horticulture:

Diversification of Crops:

Agroecology encourages the cultivation of diverse crop species and varieties within horticultural systems. Crop diversification enhances ecological resilience, reduces the risk of pest and disease outbreaks, and improves soil health through varied root systems and nutrient requirements.

Companion Planting:

Companion planting involves growing compatible plant species together to maximize their benefits and minimize negative interactions. In horticulture, certain plant combinations can deter pests, attract beneficial insects, and improve pollination, leading to more productive and ecologically balanced crop systems.

Crop Rotation:

Crop rotation is a fundamental agroecological practice in horticulture. It involves systematically changing the location of crops in successive growing seasons to prevent the buildup of pests and diseases and maintain soil fertility. Crop rotation contributes to sustainable and resilient horticultural systems.

Soil Health Management:

Agroecology prioritizes soil health and fertility through the use of organic matter, cover crops, and reduced tillage. These practices enhance soil structure, water retention, nutrient cycling, and microbial diversity, leading to healthier and more productive horticultural crops.

Integrated Pest Management (IPM):

Agroecological horticulture relies on integrated pest management strategies, combining biological control, cultural practices, and natural pesticides to manage pests while minimizing environmental impacts. IPM ensures a balanced approach to pest control, preserving biodiversity and ecological harmony.

Biodiversity Conservation in Horticultural Crops:

Habitat Creation:

Agroecological approaches in horticulture promote the creation of diverse habitats within and around agricultural fields. Hedgerows, wildflower strips, and natural areas provide shelter and food sources for beneficial insects, birds, and other wildlife. These habitats enhance biodiversity, contributing to pest control and pollination services.

Pollinator-Friendly Practices:

Agroecological horticulture recognizes the importance of pollinators and implements practices to support their populations. By providing flowering plants, avoiding the use of harmful pesticides, and creating nesting sites, agroecological systems attract and sustain pollinators, benefiting crop pollination and seed production.

Genetic Diversity:

Maintaining genetic diversity in horticultural crops is crucial for conserving biodiversity and ensuring the resilience of plant populations. Agroecology encourages the preservation and cultivation of traditional and heirloom crop varieties, which often possess unique characteristics and adaptability to local conditions.

Conservation of Native Plant Species:

Agroecological horticulture values the conservation of native plant species, which are adapted to local ecosystems and support native wildlife. By incorporating native plants into agricultural landscapes, horticulturists contribute to biodiversity conservation and ecosystem restoration.

Agrobiodiversity Parks and Gardens:

Agroecological horticulture fosters the establishment of agrobiodiversity parks and gardens, showcasing a wide range of crop species, landraces, and wild relatives. These living gene banks serve as educational resources, research centers, and sanctuaries for endangered plant varieties.

Collaboration with Indigenous and Local Communities:

Agroecological horticulture acknowledges the traditional knowledge and practices of indigenous and local communities in conserving biodiversity. Collaborating with these communities fosters sustainable horticultural approaches that respect cultural values.

By embracing agroecological approaches and promoting biodiversity conservation of horticultural crops, we can build resilient and sustainable agricultural systems. These practices not only enhance the health of ecosystems and the diversity of plant and animal species but also support the well-being of farmers, communities, and consumers. Agroecology fosters a deeper understanding of the interconnectedness between agriculture and nature, leading to more harmonious and balanced agricultural practices. Through agroecological principles, horticulturists contribute to the conservation of biodiversity and the protection of our natural heritage for future generations.

Chapter 17:

Emerging Trends in Horticulture: Hi-Tech Horticulture - Automation and Precision Agriculture

Advancements in technology are revolutionizing the field of horticulture, leading to the emergence of Hi-Tech Horticulture. This chapter explores the application of automation and precision agriculture in horticultural practices. These cutting-edge technologies enhance efficiency, productivity, and sustainability, transforming the way crops are grown and managed.

Automation in Horticulture:

Robotic Farming:

Robotic farming involves the use of autonomous robots to perform various horticultural tasks. These robots can plant seeds, apply fertilizers and pesticides, monitor crop health, and even perform harvesting. They operate based on pre-programmed algorithms or real-time data analysis, reducing the need for manual labor and increasing precision in crop management. Ex. The "LettuceBot" is a robotic weeder developed by Blue River Technology (now John Deere). It uses computer vision and machine learning algorithms to identify and selectively spray herbicides on unwanted weeds, reducing chemical usage and minimizing environmental impact.

Automated Greenhouses:

Automated greenhouses incorporate sensors, actuators, and control systems to optimize environmental conditions for crop growth. These systems regulate temperature, humidity, light intensity, and irrigation, creating a controlled environment that maximizes plant growth and minimizes resource waste. Ex. The Priva Climate Optimizer is an automated greenhouse climate control system that monitors and adjusts temperature, humidity, and CO_2 levels. It uses real-time data to create the ideal growing conditions for various horticultural crops, improving crop quality and yield.

Drone Technology:

Drones equipped with multispectral cameras and sensors are employed in horticulture for crop monitoring and analysis. They provide valuable insights into plant health, nutrient deficiencies, and pest infestations. This data allows farmers to make informed decisions and implement targeted interventions. Example: DroneDeploy is a drone mapping and analytics platform that offers vegetation indices and thermal maps for precision agriculture. Farmers can identify stress areas in the field, optimize irrigation, and detect early signs of disease using drone data.

Precision Agriculture in Horticulture:

Variable Rate Technology (VRT):

Precision agriculture uses VRT to apply inputs such as fertilizers, water, and pesticides at variable rates based on real-time data. VRT optimizes resource utilization, reduces environmental impact, and ensures that crops receive the right amount of nutrients and water, precisely where they need it. Example: Soil testing combined with VRT technology allows farmers to create site-specific nutrient application maps. The system adjusts the rate of fertilizer application based on soil nutrient levels, leading to efficient nutrient management and improved crop yield.

GPS Guidance Systems:

GPS-guided machinery enables precise control of planting, spraying, and harvesting operations. These systems enhance field efficiency, reduce overlap, and minimize soil compaction.

Example: The Trimble EZ-Pilot is a GPS steering system that can be retrofitted onto existing agricultural machinery. It provides accurate guidance, reducing operator fatigue and increasing operational efficiency.

Remote Sensing and Internet of Things (IoT):

Remote sensing technologies, combined with IoT devices, allow continuous monitoring of crop conditions and environmental factors. Data collected from sensors on the field is transmitted to central systems, enabling real-time decision-making and proactive management.

Example: Smart irrigation systems use soil moisture sensors and weather data to adjust irrigation schedules and water delivery. This ensures that crops receive the right amount of water, reducing water wastage and optimizing water use efficiency.

Data Analytics and Artificial Intelligence (AI):

Data analytics and AI are utilized to process vast amounts of data collected from various sources, including sensors, drones, and satellites. AI algorithms analyze this data to generate valuable insights, predictive models, and personalized recommendations for horticultural practices.

Example: The Climate FieldView platform from The Climate Corporation uses AI to provide data-driven insights and prescriptions for farmers. It helps optimize planting density, monitor crop health, and predict yield potential based on historical and real-time data.

In conclusion, Hi-Tech Horticulture through automation and precision agriculture is transforming the horticultural industry, making it more efficient, sustainable, and data-driven. By integrating cutting-edge technologies such as robotics, automation, drones, and AI into horticultural practices, farmers can enhance crop management, improve resource utilization, and reduce environmental impact. Hi-Tech Horticulture not only boosts productivity and profitability but also enables the cultivation of crops in a more environmentally friendly and socially responsible manner. As these technologies continue to evolve and become more accessible, they hold the potential to revolutionize global horticultural practices and address the challenges of feeding a growing population sustainably.

Chapter 18:

Genetic Engineering and Genome Editing in Horticulture

Genetic engineering and genome editing are powerful biotechnological tools that have revolutionized the field of horticulture. This chapter explores how these techniques are applied in horticultural crops, their benefits, and the ethical considerations involved.

Genetic Engineering in Horticulture:

Definition and Process:

Genetic engineering involves the manipulation of an organism's genetic material to introduce new traits or modify existing ones. In horticulture, this is typically done by inserting specific genes from other organisms (often unrelated species) into the target plant's genome. The transferred genes can confer desirable traits, such as pest resistance, disease tolerance, or improved nutritional content.

Benefits of Genetic Engineering in Horticulture:

➢ Pest and Disease Resistance:

Genetic engineering has enabled the development of horticultural crops that are resistant to pests and diseases. For example, Bt (Bacillus thuringiensis) genes have been inserted into crops like cotton and corn to produce insecticidal proteins, reducing the need for chemical pesticides.

➢ Enhanced Nutritional Content:

Genetic engineering has been used to increase the nutritional content of horticultural crops. For instance, Golden Rice, a genetically modified rice variety, contains higher levels of vitamin A, addressing micronutrient deficiencies in populations reliant on rice as a staple food.

➢ Extended Shelf Life:

Some genetically engineered horticultural crops have improved post-harvest characteristics, such as delayed ripening or reduced spoilage. This leads to an extended shelf life, reducing food waste and improving supply chain efficiency.

➢ Abiotic Stress Tolerance:

Genetic engineering has allowed the development of horticultural crops with enhanced tolerance to environmental stresses, such as drought, salinity, and extreme temperatures. These traits contribute to increased resilience and yield stability.

- Example of Genetic Engineering in Horticulture:

Flavr Savr Tomato: The Flavr Savr tomato was one of the first genetically engineered horticultural crops approved for commercial production. It was engineered to delay softening and ripening, leading to a longer shelf life and improved transportation. Although the Flavr Savr tomato was not commercially successful, it laid the foundation for subsequent genetic engineering efforts in horticulture.

Genome Editing in Horticulture:

Definition and Process:

Genome editing is a precise and targeted modification of an organism's DNA, allowing for specific changes in the genome without the introduction of foreign genes. The most commonly used genome editing technique is CRISPR-Cas9, which acts like a pair of molecular scissors, enabling precise cutting and editing of DNA sequences.

Benefits of Genome Editing in Horticulture:

- ➢ Gene Knockout and Silencing:

Genome editing allows the knockout or silencing of specific genes responsible for undesirable traits. This approach is valuable for improving fruit quality, modifying flowering times, and enhancing other agronomically important traits.

- ➢ Gene Editing for Disease Resistance:

Through genome editing, horticultural crops can be engineered to have enhanced resistance against viral, bacterial, and fungal pathogens. By disabling susceptibility genes or introducing specific resistance genes, crops can be protected from infectious diseases.

- ➢ Precision Trait Modification:

Genome editing enables precise modifications to specific regions of the genome, allowing for the fine-tuning of desirable traits without introducing unwanted changes.

- ➢ Reduction of Off-Target Effects:

Recent advancements in genome editing techniques have reduced off-target effects, making the process more precise and efficient.

- Example of Genome Editing in Horticulture:

Seedless Watermelon: Using CRISPR-Cas9, scientists have successfully edited genes involved in seed development in watermelon, resulting in the creation of seedless varieties. This genetic modification eliminates the need for manual seed removal, streamlining the harvesting process and improving consumer experience.

Ethical Considerations:

Genetic engineering and genome editing in horticulture raise several ethical considerations, including concerns about the potential environmental impact, unintended consequences, and consumer acceptance. The release of genetically modified organisms (GMOs) into the environment requires rigorous safety assessments to ensure they do not harm native ecosystems or non-target organisms. Transparency in labeling GMO products is also crucial to allow consumers to make informed choices.

In conclusion, genetic engineering and genome editing have ushered in a new era of innovation and possibilities in horticulture. These technologies hold the potential to address pressing challenges such as food security, nutritional enhancement, and sustainable agriculture. However, their deployment requires careful consideration of ethical, environmental, and regulatory aspects to ensure that the benefits of these biotechnologies are harnessed responsibly for the betterment of society and the environment.

Chapter 19:

Biophilic Design and Therapeutic Horticulture

Biophilic design and therapeutic horticulture are two interconnected approaches that harness the innate human connection with nature to promote well-being and create harmonious environments. Biophilic design integrates natural elements into indoor spaces, such as living walls, natural light, and water features, to improve productivity, reduce stress, and enhance cognitive function. It fosters a sense of tranquility and connection with the natural world. Therapeutic horticulture, on the other hand, uses gardening and plant-related activities to improve physical, emotional, and mental health.

Biophilic Design:

Definition and Principles:

Biophilic design is a design philosophy that seeks to incorporate nature and natural elements into the built environment. It is based on the idea that humans have an innate affinity for nature and benefit from having access to green spaces, natural light, and views of nature. Biophilic design aims to create spaces that foster a sense of connection with nature, promoting health, happiness, and productivity.

Principles of Biophilic Design:

Natural Light and Views:

Incorporating ample natural light and providing views of nature from indoor spaces are essential elements of biophilic design. Exposure to natural light is known to improve mood and productivity, while views of nature can reduce stress and enhance cognitive function.

Natural Materials and Patterns:

Using natural materials, such as wood, stone, and plants, and incorporating organic patterns in design elements contribute to a biophilic environment. These materials and patterns evoke a sense of nature and create a soothing and comforting ambiance.

Indoor Plants and Living Walls:

Including indoor plants and living walls in interior spaces brings the benefits of nature indoors. Plants not only improve air quality but also add aesthetic appeal and create a connection with nature.

Water Features:

Water features, such as fountains or indoor ponds, can mimic natural water bodies and promote a sense of tranquility and relaxation.

Example of Biophilic Design:

Amazon Spheres: The Amazon Spheres in Seattle, USA, are an example of biophilic design at a grand scale. These three interconnected glass domes house more than 40,000 plants from around the world. The lush greenery, natural light, and diverse plant life create an immersive and biophilic environment for Amazon employees to work and relax.

Therapeutic Horticulture:

Definition and Benefits:

Therapeutic horticulture is the use of horticultural activities and gardening as a therapeutic tool to improve physical, mental, and emotional well-being. Engaging in gardening and horticultural activities can have positive effects on stress reduction, mood enhancement, and overall health.

Therapeutic Horticulture Programs:

Healing Gardens:

Healing gardens are designed to provide a calming and restorative environment for patients, staff, and visitors in healthcare settings. These gardens often include sensory elements such as fragrant flowers, soothing water features, and comfortable seating areas.

Horticultural Therapy:

Horticultural therapy involves structured and guided horticultural activities to achieve specific therapeutic goals. It is used as a complementary treatment in rehabilitation, mental health, and special needs programs.

Community Gardens:

Community gardens provide opportunities for social interaction and community engagement. Gardening in a group setting fosters a sense of belonging and connection with others, promoting social well-being.

Memory Gardens:

Memory gardens are designed for individuals with memory-related conditions such as Alzheimer's disease. These gardens incorporate familiar and sensory elements to stimulate memories and provide a safe and comforting space for individuals and their caregivers.

Example of Therapeutic Horticulture:

One notable example of therapeutic horticulture in India is the use of gardening and horticultural activities in various rehabilitation centers and mental health institutions. These centers incorporate gardening and plant-related activities into their therapeutic programs to provide a healing and empowering experience for individuals with mental health conditions, such as depression, anxiety, and post-traumatic stress disorder.

In conclusion, biophilic design and therapeutic horticulture highlight the intrinsic relationship between humans and nature and demonstrate how incorporating nature into the built environment can have profound effects on our well-being. Biophilic design creates spaces that nurture our connection with nature, promoting a sense of calm, comfort, and vitality. Therapeutic horticulture, on the other hand, uses gardening and horticultural activities as powerful tools for healing and improving the physical, emotional, and social aspects of our lives. By embracing these concepts, we can create environments that support human health, happiness, and resilience, fostering a deeper and more meaningful relationship with nature.

Chapter 20

Horticultural Extension and Outreach

Horticultural extension and outreach play a vital role in disseminating knowledge, skills, and technologies to farmers and communities involved in horticultural practices. This chapter explores the significance of extension services in horticulture, the implementation of farmer field schools, and the social impact of community gardens.

Role of Extension in Horticulture:

Definition and Objectives:

Horticultural extension is a system that connects research, knowledge, and best practices from experts to farmers and horticulturists. The main objective of horticultural extension is to improve the adoption of modern and sustainable horticultural practices, enhance crop productivity, and promote the well-being of farming communities.

Services Provided by Extension Agents:

Extension agents, often working under government agencies, NGOs, or agricultural universities, offer a range of services to farmers and horticulturists. These include:

- Conducting on-farm demonstrations and trials to showcase improved horticultural practices.
- Providing technical advice on crop management, pest and disease control, and post-harvest handling.
- Organizing training workshops and capacity-building programs for farmers.
- Facilitating access to improved seeds, planting materials, and other inputs.
- Promoting the adoption of sustainable and climate-smart practices.
- Assisting with market linkages and value addition to horticultural produce.

Example of Horticultural Extension:

The "Krishi Vigyan Kendra" (Agricultural Science Center) program in India is an exemplary horticultural extension initiative. These centers, established by the Indian Council of Agricultural Research (ICAR), provide agricultural and horticultural expertise to farmers at the grassroots level. Extension agents at these centers conduct training programs, field demonstrations, and technology dissemination activities to improve agricultural and horticultural practices among smallholder farmers.

Farmer Field Schools and Knowledge Transfer:

Definition and Approach:

Farmer Field Schools (FFS) are participatory and experiential learning platforms where farmers come together to acquire new skills and share knowledge with each other. FFS emphasize learning-by-doing, empowering farmers to make informed decisions based on their experiences and observations.

Components of Farmer Field Schools:

- Season-long practical training on horticultural practices, including crop production, pest and disease management, and post-harvest handling.
- Group discussions, knowledge exchange, and problem-solving sessions among farmers.
- Hands-on field activities, such as planting, pruning, and monitoring crop growth.

Benefits of Farmer Field Schools:

- Enhanced farmer empowerment and decision-making abilities.
- Improved understanding of sustainable horticultural practices.
- Strengthened social cohesion and community support among farmers.
- Adoption of best practices leading to increased crop productivity and income.

Example of Farmer Field Schools:

The Integrated Pest Management (IPM) Farmer Field School approach is widely used in horticulture to promote sustainable pest management practices. Farmers are trained on recognizing pests, natural enemies, and pest damage. They learn how to implement ecologically sound pest control measures, such as the use of biopesticides and trap crops, reducing their reliance on synthetic chemical pesticides.

Community Gardens and Social Impact:

Definition and Purpose:

Community gardens are shared spaces where individuals or groups from the local community collectively grow fruits, vegetables, and ornamental plants. These gardens serve multiple purposes, including food production, education, recreation, and community building.

Social Impact of Community Gardens:

- Social Cohesion: Community gardens foster a sense of belonging and community ownership, as people come together to cultivate the land and share in the harvest.
- Education and Skill Development: Community gardens provide opportunities for individuals to learn about horticulture, gardening, and sustainable agriculture practices.
- Food Security: Community gardens can contribute to food security by providing fresh, locally grown produce to community members, especially in urban areas with limited access to fresh fruits and vegetables.
- Health and Well-being: Engaging in gardening activities and spending time in green spaces have been linked to improved mental health, reduced stress, and enhanced overall well-being.

Example of Community Gardens:

The "Incredible Edible" movement in the United Kingdom is a successful example of community gardens making a positive impact. It started in the town of Todmorden, where residents began planting food crops in public spaces. Today, the movement has spread to various cities across the UK, promoting community engagement, local food production, and sustainability.

In conclusion, horticultural extension and outreach are crucial components of agricultural development, helping farmers and communities adopt best practices, improve productivity, and enhance their livelihoods. Farmer Field Schools facilitate hands-on learning and knowledge exchange, empowering farmers to make informed decisions and improve their horticultural practices. Community gardens serve as valuable spaces for social interaction, education, and food production, making positive contributions to the health and well-being of communities. Through these initiatives, horticultural extension and outreach play a significant role in promoting sustainable horticultural practices and fostering resilient and vibrant agricultural communities.

Future Prospects and Challenges in Horticulture

Horticulture faces a dynamic landscape of challenges and opportunities in the coming years. This chapter explores the global challenges and opportunities in horticulture, with a specific focus on climate change and adaptation strategies. Additionally, it discusses the policy implications and role of horticulture in achieving Sustainable Development Goals (SDGs).

Global Challenges and Opportunities in Horticulture:

> Population Growth and Food Security:

The world's population is projected to reach 9.7 billion by 2050, leading to increased demand for food, including horticultural products. Horticulture presents an opportunity to enhance food security through diversified and nutritious crops that can be grown in a range of environments.

> Resource Scarcity and Sustainable Practices:

As natural resources such as water and arable land become scarce, horticulture must adopt sustainable practices to ensure long-term productivity. Efficient water management, soil conservation, and reduced waste in the supply chain are crucial considerations.

> Urbanization and Vertical Farming:

Rapid urbanization creates challenges for traditional agricultural practices. Urban horticulture, including vertical farming and rooftop gardens, offers innovative solutions to produce fresh food in urban areas, reduce food miles, and create green spaces.

> Biodiversity Loss and Genetic Resources:

The loss of biodiversity poses a threat to horticultural crops and their resilience to changing conditions. Preserving and utilizing genetic resources in horticulture can contribute to crop improvement and adaptation.

> Market Access and Value Addition:

Access to markets, fair prices, and value addition are crucial for horticulturists to thrive economically. Strengthening supply chains, improving post-harvest handling, and promoting niche markets can enhance the competitiveness of horticultural products.

Climate Change and Adaptation Strategies:

Impact of Climate Change on Horticulture:

Climate change poses significant challenges to horticulture, affecting crop growth, yields, and quality. Shifts in temperature, precipitation patterns, and extreme weather events can lead to increased pest and disease pressure and alter crop phenology.

Adaptation Strategies in Horticulture:

➤ Crop Diversification:

Diversifying horticultural crops can enhance resilience to climate variability. Farmers can explore growing new crop varieties or adopting underutilized indigenous species that are better adapted to changing conditions.

➤ Water Management:

Improved water management practices, such as drip irrigation and rainwater harvesting, can optimize water use efficiency and mitigate the impact of water scarcity on horticultural crops.

➤ Heat and Cold Tolerance:

Breeding horticultural crops for heat and cold tolerance can enable them to withstand extreme temperatures and maintain productivity under adverse climate conditions.

➤ Protected Cultivation:

Using greenhouses and other protected cultivation methods provides a controlled environment that shields horticultural crops from extreme weather events and allows for year-round production.

➤ Climate Information Services:

Access to timely and accurate climate information is crucial for horticulturists to plan and make informed decisions. Climate information services can provide farmers with weather forecasts, climate predictions, and advisory services.

> Example of Adaptation Strategies:

The "Climate-Resilient Agribusiness for Tomorrow" (CRAFT) project in Bangladesh supports horticulturists to adapt to climate change. The project focuses on climate-resilient horticultural practices, capacity-building for farmers, and promoting climate-smart technologies such as high-efficiency irrigation and weather-resistant crop varieties.

Policy Implications and Sustainable Development Goals:

> Horticulture and Sustainable Development Goals (SDGs):

Horticulture plays a significant role in achieving several SDGs, including Goal 1 (No Poverty) by providing income opportunities to smallholder farmers, Goal 2 (Zero Hunger) through improved food security, and Goal 3 (Good Health and Well-Being) by promoting a diverse and nutritious diet.

> Policy Support for Sustainable Horticulture:

Government policies and incentives can promote sustainable horticultural practices. Policies that support research and innovation, invest in agricultural infrastructure, and provide market linkages can create an enabling environment for horticultural development.

> Inclusive and Gender-Sensitive Policies:

Policies should be inclusive and gender-sensitive, recognizing the contributions of women in horticulture. Access to resources, credit, and training opportunities should be equitable, empowering women farmers and entrepreneurs.

> Climate Change Policy Integration:

Integration of climate change considerations into agricultural policies is essential. Policies that encourage climate-smart practices and incentivize climate resilience can enhance horticultural adaptation to changing climate conditions.

> Example of Policy Integration:

The "Climate Smart Agriculture and Rural Enterprise Programme" (SAFER) in Kenya integrates climate-smart practices into horticulture value chains. The program promotes sustainable land management, water use efficiency, and improved access to markets, supporting the livelihoods of smallholder farmers.

In conclusion, the future of horticulture is intertwined with the global challenges and opportunities it faces. Climate change poses a significant threat, necessitating adaptation strategies to ensure horticultural resilience. Embracing sustainable practices, fostering innovation, and integrating climate change considerations into policies are essential for the growth and development of horticulture. The role of horticulture in achieving Sustainable Development Goals is pivotal, as it addresses food security, nutrition, poverty reduction, and sustainable livelihoods. By recognizing the challenges and leveraging opportunities, horticulture can contribute to a resilient and sustainable future for agriculture and humanity.

The Vision for a Sustainable and Resilient Horticultural Sector

The horticultural sector plays a vital role in food production, environmental conservation, and socio-economic development. As we face global challenges such as climate change, resource scarcity, and food insecurity, it is crucial to envision a sustainable and resilient horticultural sector. This chapter outlines the vision and key principles for achieving a sustainable and resilient horticultural future.

Vision for a Sustainable Horticultural Sector:

The vision for a sustainable horticultural sector is one that balances environmental stewardship, social equity, and economic prosperity. It envisions a world where horticultural practices safeguard natural resources, protect biodiversity, and mitigate climate change impacts while ensuring food security, livelihoods, and improved well-being for horticulturists and communities.

Key Principles for a Sustainable Horticultural Sector:

 ➢ Climate Smart and Resilient Practices:

The horticultural sector must embrace climate-smart practices that enhance resilience to climate change. This includes adopting drought-tolerant crop varieties, implementing water-efficient irrigation systems, and promoting protected cultivation to mitigate extreme weather events.

 ➢ Biodiversity Conservation:

Preserving and utilizing genetic diversity in horticultural crops is essential for resilience and adaptation. Efforts to conserve heirloom varieties, wild relatives, and indigenous species contribute to crop improvement and promote ecosystem health.

 ➢ Sustainable Resource Management:

Horticulture should prioritize sustainable resource management, including water, soil, and energy. Implementing efficient water use, soil conservation practices, and renewable energy solutions contribute to a low-carbon and resource-efficient sector.

➢ Inclusive and Empowering Practices:

The horticultural sector should promote inclusivity and gender equity, recognizing the contributions of women and marginalized groups. Equitable access to resources, knowledge, and decision-making opportunities fosters sustainable and socially just horticultural practices.

➢ Policy Support and Collaborative Partnerships:

Government policies that prioritize sustainable horticulture and incentivize environmentally friendly practices are critical. Public-private partnerships and collaboration between stakeholders can amplify the impact of sustainable initiatives.

Role of Education and Research:

Education and research are essential pillars for achieving a sustainable and resilient horticultural sector. Investment in horticultural education and research institutions fosters innovation, advances technology adoption, and addresses emerging challenges.

➢ Circular Economy and Zero Waste:

The horticultural sector can embrace circular economy principles, minimizing waste and promoting the reuse and recycling of organic matter and by-products. Composting, bioenergy generation, and waste reduction strategies contribute to a zero-waste approach.

➢ Knowledge Sharing and Capacity Building:

Continual learning and knowledge exchange are essential for a sustainable horticultural sector. Providing access to technical expertise, farmer training programs, and extension services enhances the adoption of best practices.

➢ Market Linkages and Value Addition:

Strengthening market linkages and value addition opportunities ensures horticultural products reach consumers efficiently and sustainably. Promotion of niche markets, organic certification, and fair trade practices contribute to economic viability for horticulturists.

Importance of Consumer Awareness:

Consumer awareness and demand for sustainably produced horticultural products drive market transformation. Education campaigns and eco-labeling initiatives empower consumers to make informed choices and support sustainable horticulture.

Embracing Technological Advancements:

Technological advancements such as precision agriculture, remote sensing, and blockchain can enhance the efficiency and transparency of horticultural practices. Embracing these innovations can optimize resource use and traceability in the supply chain.

In conclusion, the vision for a sustainable and resilient horticultural sector envisions a future where horticulture thrives while safeguarding natural resources and promoting social well-being. By embracing climate-smart practices, conserving biodiversity, and promoting sustainable resource management, the horticultural sector can navigate global challenges and seize opportunities for growth. Education, research, consumer awareness, and collaboration are instrumental in achieving this vision, as they foster innovation, support informed decision-making, and drive market transformation. By adhering to the key principles outlined in this chapter, horticulture can be a driving force in building a more sustainable and resilient future for agriculture and humanity.

"Nurture Nature's Secrets: 50 Essential One-Liners"

1. What is horticulture?

   Horticulture is the science and art of growing and cultivating plants for food, medicine, ornamentation, and recreation.

2. What are the main sub-disciplines of horticulture?

   The main sub-disciplines of horticulture include pomology, olericulture, floriculture, landscape horticulture, and medicinal horticulture.

3. What is the significance of horticulture in society?

   Horticulture plays a vital role in providing food, beautifying landscapes, improving the environment, and contributing to the economy.

4. How is horticulture different from agriculture?

   Horticulture focuses on growing fruits, vegetables, and ornamental plants, while agriculture encompasses broader practices, including large-scale cultivation of crops and livestock.

5. What is the importance of soil in horticulture?

   Soil provides essential nutrients, water, and support for plant growth, making it a fundamental aspect of horticulture.

6. How does photosynthesis benefit horticultural plants?

   Photosynthesis enables plants to convert sunlight into energy, supporting their growth and development.

7. What are the primary components of a flower?

   Flowers consist of petals, sepals, stamens, and pistils, with each playing a role in plant reproduction.

8. How is a greenhouse beneficial in horticulture?

   Greenhouses provide a controlled environment, allowing year-round cultivation and protection of plants from adverse weather conditions.

9. What is the purpose of pruning in horticulture?

   Pruning helps shape plants, remove dead or diseased parts, and promote better growth and flowering.

10. How do plants take up water from the soil?

Plants absorb water through their roots and transport it through their vascular system.

11. What are the two main types of plant propagation?

The two main types of plant propagation are sexual (by seeds) and asexual (by vegetative means).

12. How can farmers manage pests in horticulture sustainably?

Integrated Pest Management (IPM) combines multiple approaches, such as biological control and cultural practices, to control pests effectively.

13. What are the benefits of crop rotation in horticulture?

Crop rotation helps prevent soil depletion, reduces pest and disease buildup, and improves overall crop health.

14. How does cross-pollination occur in horticultural crops?

Cross-pollination happens when pollen from one flower is transferred to the stigma of another flower of the same species.

15. What is the role of hormones in plant growth?

Hormones regulate plant growth and development processes, such as germination, flowering, and fruit development.

16. What are the main nutrients essential for plant growth?

The main nutrients required by plants are nitrogen, phosphorus, and potassium, often represented by the NPK ratio.

17. How can hydroponics benefit horticulture?

Hydroponics enables soilless cultivation, conserves water, and promotes faster plant growth.

18. Why is sustainable horticulture essential for the environment?

Sustainable horticulture practices protect natural resources, reduce environmental impact, and promote long-term productivity.

19. What is the primary method of seed propagation in horticulture?

Seed propagation is the most common method of reproducing horticultural crops, offering genetic diversity.

20. How can horticulture promote biodiversity conservation?

Growing diverse plant species in gardens and landscapes can create habitats for various wildlife, contributing to biodiversity conservation.

21. What are the different types of soil erosion and their prevention methods?

Types of soil erosion include water erosion, wind erosion, and sheet erosion. Prevention methods include terracing, windbreaks, and cover crops.

22. How do biotechnological approaches benefit plant breeding in horticulture?

Biotechnological methods, such as genetic engineering and genome editing, help develop crops with improved traits and resistance to diseases.

23. How does urban horticulture contribute to sustainable living?

Urban horticulture promotes green spaces, rooftop gardens, and urban farming, enhancing air quality and food availability in cities.

24. What are the basic principles of sustainable horticulture?

Sustainable horticulture emphasizes environmental stewardship, social equity, and economic viability in agricultural practices.

25. How does horticulture promote human well-being and mental health?

Therapeutic horticulture uses gardening as a means of therapy, providing relaxation and improving mental well-being.

26. What are the benefits of vertical gardening in urban environments?

Vertical gardening maximizes space utilization, making it suitable for small urban areas and promoting green infrastructure.

27. How can horticulture contribute to food security?

Horticulture provides a diverse range of fruits, vegetables, and other crops, increasing dietary variety and supporting food security.

28. What is the role of community gardens in urban settings?

Community gardens foster social cohesion, promote sustainable practices, and enhance community well-being through shared gardening experiences.

29. How does agroecology influence horticultural practices?

Agroecological approaches integrate ecological principles into horticultural systems, emphasizing biodiversity, nutrient cycling, and ecological balance.

30. How can precision agriculture improve horticultural productivity?

Precision agriculture uses technology to optimize resource use, improve crop monitoring, and increase efficiency in horticulture.

31. What are the challenges of genetic engineering in horticulture?

Challenges include public perception, potential environmental impacts, and ensuring ethical use of genetic modification.

32. What is the concept of biophilic design in horticulture?

Biophilic design incorporates natural elements into indoor spaces to improve human well-being and connection with nature.

33. How does nursery management impact horticultural crop production?

Proper nursery management ensures the production of healthy seedlings, improving overall crop success and yields.

34. What are the benefits of rooftop gardening and balcony gardens?

Rooftop gardening and balcony gardens utilize urban spaces for greenery, enhancing aesthetics and providing food sources in cities.

35. How can horticulture be adapted to climate change challenges?

Implementing climate-resilient horticultural practices, such as drought-tolerant crops and efficient irrigation, helps mitigate climate change impacts.

36. What are the Sustainable Development Goals (SDGs) related to horticulture?

SDGs, such as Zero Hunger, Climate Action, and Life on Land, align with horticulture's role in achieving global sustainability.

37. What is the significance of post-harvest management in horticulture?

Post-harvest management ensures the preservation and quality of horticultural produce after harvesting, reducing food waste and losses.

38. How can horticultural extension services benefit farmers?

Horticultural extension services provide knowledge, technical support, and best practices to improve farmers' productivity and income.

39. What are the environmental benefits of using organic horticultural practices?

Organic horticultural practices promote soil health, reduce chemical pollution, and protect natural ecosystems.

40. How can horticulture contribute to biodiversity conservation?

Growing diverse plant species in gardens and landscapes can create habitats for various wildlife, contributing to biodiversity conservation.

41. How can horticulture promote urban greening and sustainable cities?

Urban greening initiatives, such as street trees and public parks, improve air quality and create healthier and more livable urban environments.

42. How does biophilic design in horticulture impact human well-being?

Biophilic design enhances mental health, reduces stress, and improves cognitive function by integrating nature into built environments.

43. What are the common pests and diseases in horticulture?

Common pests include aphids, caterpillars, and mites, while common diseases include powdery mildew, blight, and rust.

44. How can horticulture contribute to the restoration of degraded landscapes?

Horticultural practices, such as afforestation and reforestation, play a vital role in restoring degraded lands and improving ecosystems.

45. What are the benefits of community gardens in urban settings?

Community gardens foster social cohesion, promote sustainable practices, and enhance community well-being through shared gardening experiences.

46. How does agroecology influence horticultural practices?

Agroecological approaches integrate ecological principles into horticultural systems, emphasizing biodiversity, nutrient cycling, and ecological balance.

47. What is the concept of biophilic design in horticulture?

Biophilic design incorporates natural elements into indoor spaces to improve human well-being and connection with nature.

48. How does nursery management impact horticultural crop production?

Proper nursery management ensures the production of healthy seedlings, improving overall crop success and yields.

49. What are the benefits of rooftop gardening and balcony gardens?

Rooftop gardening and balcony gardens utilize urban spaces for greenery, enhancing aesthetics and providing food sources in cities.

50. How can horticulture be adapted to climate change challenges?

Implementing climate-resilient horticultural practices, such as drought-tolerant crops and efficient irrigation, helps mitigate climate change impacts.

Glossary

1. Horticulture: The science and art of cultivating plants, including fruits, vegetables, flowers, ornamental plants, and medicinal herbs, for human use and enjoyment.

2. Botany: The scientific study of plants, including their structure, growth, reproduction, and classification.

3. Photosynthesis: The process by which green plants and some other organisms convert light energy into chemical energy, producing glucose and oxygen from carbon dioxide and water.

4. Transpiration: The process by which plants release water vapor through their leaves into the atmosphere.

5. Propagation: The process of creating new plants from seeds, cuttings, grafts, or other vegetative means.

6. Pollination: The transfer of pollen from the male reproductive organ (anther) to the female reproductive organ (stigma) of a flower, leading to fertilization and seed formation.

7. Fertilizer: A substance containing essential nutrients (such as nitrogen, phosphorus, and potassium) applied to soil or plants to improve growth and productivity.

8. Integrated Pest Management (IPM): A sustainable approach to managing pests that combines biological, cultural, physical, and chemical control methods to minimize their impact while safeguarding the environment and human health.

9. Grafting: A method of vegetative propagation where a shoot or bud (scion) from one plant is inserted onto the stem or root of another plant (rootstock) to grow as a single plant.

10. Hybrid: The offspring resulting from the crossbreeding of two genetically different plants or plant varieties.

11. Pruning: The act of trimming or cutting back branches, shoots, or roots of a plant to promote growth, improve shape, and maintain plant health.

12. Perennial: A plant that lives for more than two years, typically flowering and producing seeds multiple times during its lifespan.

13. Annual: A plant that completes its entire life cycle, from germination to seed production, within one growing season.

14. Biennial: A plant that takes two growing seasons to complete its life cycle, typically flowering and producing seeds in its second year.

15. Rhizome: An underground stem from which new shoots and roots grow, enabling vegetative propagation in plants like ginger and iris.

16. Xeriscaping: A landscaping technique that focuses on using drought-tolerant plants and minimizing water consumption in arid or water-limited regions.

17. Greenhouse: A controlled environment structure made of glass or plastic, used for growing plants in controlled conditions, extending the growing season, and protecting plants from adverse weather.

18. Hydroponics: A method of growing plants in a nutrient-rich water solution without soil, providing essential nutrients directly to the plant roots.

19. Compost: Decomposed organic matter used as a soil amendment to enrich soil fertility and improve its structure.

20. Hardiness Zone: A geographical area characterized by its average annual minimum temperature, used to determine the suitability of plants for specific regions.

21. Micropropagation: A tissue culture technique that involves growing plant cells or tissues in a controlled environment to produce large quantities of identical plants.

22. Drought Resistance: The ability of a plant to withstand extended periods of low water availability without significant harm.

23. Allelopathy: The phenomenon where certain plants release chemicals that inhibit the growth of neighboring plants, impacting plant competition and growth.

24. Sustainable Agriculture: Farming practices that aim to meet current food production needs while preserving natural resources and maintaining ecological balance for future generations.

25. Pest Resistance: The ability of a plant to withstand damage caused by pests or pathogens through genetic traits or other defense mechanisms.

This glossary provides a comprehensive list of essential horticultural terms to help readers grasp the concepts and practices discussed in the book effectively.

Gardening is the purest of human pleasures." - Francis Bacon

Printed in Great Britain
by Amazon

28573039R00064